Chemistry Crosswords

Compiled by

Paul Board

Print ISBN: 978-1-78262-890-3

A catalogue record for this book is available from the British Library

Published by The Royal Society of Chemistry,
Thomas Graham House, Science Park, Milton Road,
Cambridge CB4 0WF, UK

Registered Charity Number 207890

Visit our website at www.rsc.org/books

Printed in the United Kingdom by CPI Group (UK) Ltd, Croydon, CR0 4YY, UK

Introduction

Welcome to the *Chemistry World* crossword compendium. A concoction of cryptic clues, anagrams and lateral thinking challenges, each crossword is themed around science, with some cultural conundrums thrown in for good measure. You will find nearly 70 crosswords and their solutions within this book, all meticulously constructed by our regular compiler Paul Board.

Our appreciation and thanks go to Paul for producing these enjoyable, challenging, and sometimes fiendish, cryptic crosswords. He has never failed to surprise us or our readers with his inventive and inspired clues over the years.

We hope you enjoy dusting off those chemistry books, scrubbing up on your general knowledge, and putting yourself in Paul's shoes, as you make your way through this book.

The *Chemistry World* team.

Crosswords

Crossword 1

Across

1 Later bus turns North for officer (9)
6 German chemist confused beggars on loss of strontium (5)
9 Lack of rhenium in reagent representative? (5)
10 Digitalis doctor in need of a tonic himself? (9)
11 Old photo of a greedy troupe? Maybe (13)
14 Compound sounding like it's OK after sunset (7)
16 Gift for now (7)
17 Best time of day to do the ironing? (7)
19 Requiring baking skill? Sounds like it! (7)
21 Longsightedness induced by theorem on crumpled papyri? (13)
24 Solvent causing mum unrest? (9)
26 I turn nice for ancient tribe (5)
27 Seasoned for a sailor? (5)
28 Person from Gdansk at 90 degrees? (5,4)

Down

1 Mistakenly gained trust of this polymer chemist after loss of front teeth? (10)
2 For example, brute intoxicated could carry this (4,3)
3 French and English definite articles will keep turning! (5)
4 Award ledger to chemist composer (6,5)
5 Catch what's left (3)
6 Teeny lace creation for simplest alkyne (9)
7 Effeminate when epicentre loses extreme right (7)
8 A gig returns for something big (4)
12 Hemp pioneer, scrambled, destroyer of worlds (11)
13 Shame when swirling gas emits it (10)
15 State peculiar to an individual, such as happy idiot, lost and lacking phosphorus (9)
18 Bethan, older has spirit (7)
20 In full Ancient Roman style (2,5)
22 State of rare ichthyosaur in Germany? (5)
23 Spam stew? Very current! (4)
25 Mourn loss of top old man for something to keep his ashes in (3)

Crossword 2

Across

4 Biochemist takes amine with selenium deficiency (8)
8 Larch I chopped has handedness (6)
9 Polymer found in jumble sale-neat! (8)
10 Flustered Ada takes dog or man of moles (8)
11 Respite that stands out? (6)
12 Pepper Leo lost in Japan (8)
13 Train got derailed: now spinning! (8)
16 Ask Harovian about nuclear physicist after Ian's departure (8)
19 Prince's car got something slow and slithery inside (8)
21 Ant acid not antacid (6)
23 Beautiful French castle to scale? (8)
24 Massage nice oils for breast enhancer (8)
25 Nests found in key Riesling region (6)
26 Fling wet brew for political persuasion (4-4)

Down

1 Posh avian expert has something of the playwright in him (7)
2 Walk Kermit against his will? (9)
3 Mistakenly glues end of wand for ooze? (6)
4 Sherbet drove dog crazy? Ultracentrifuge inventor appears! (7,8)
5 Halt pork preparation for German chemist (8)
6 Melt a mixture for hard rock (5)
7 Grandma eager for this fabric? (7)
14 Make habitable out of former art (9)
15 Cabbage directing Bond (8)
17 Spray 22 in A & E (7)
18 Rail sounding like request for additional poultry (7)
20 Clear top table of wine after mix up (6)
22 Most I manage is damp (5)

Crossword 3

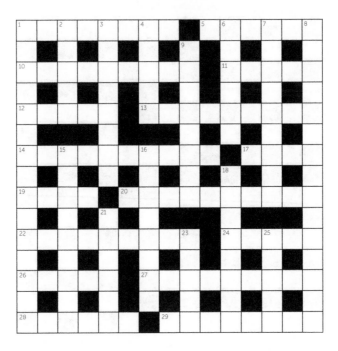

Across

1 Rearrange lime tutu for element of Paris? (8)
5 Stews egghead for candies (6)
10 Cheese for muddled men at helm? (9)
11 Prof takes oxygen for confirmation (5)
12 Country described in diary? (5)
13 Boar's feet cooked for more traditional fare (5,4)
14 Lance curbs eyesores that are also gems (10)
17 Drink to transport the dead? Sounds like it! (4)
19 Catches newts on removal of second swamp (4)
20 Litter a git hurls at the beautiful people? (10)
22 Killer at large when ex throws rice at tux? (9)
24 Fish just over five metres long?! (5)
26 Glib about the oil spill? (5)
27 Disordered Dad online for weed (9)
28 Garlic discovered on Mars (6)
29 Chaotically steering whole numbers (8)

Down

1 Turned inedible congers into something useful for the distiller (6,9)
2 Shy dimwit returned on loss of tungsten (5)
3 Metal extracted from distant alumni (8)
4 Rush around capital of Estonia for escort (5)
6 Deer returns when I tip a waiter (6)
7 Trio cease turning into mysterious things (9)
8 Mistakenly sniffs our things? Could be curtains! (4,11)
9 Battered hat leaks universal solvent (8)
15 Him untrue about Russian element (9)
16 Chore I'd left for salt after confusion (8)
18 Perplexed, preen one for rubber (8)
21 Cook copper concoction for uninvited bird (6)
23 Gas produced by returning non-executive (5)
25 Her nickel lost in river (5)

Crossword 4

Across

1 Etcher's son broke computer image (10)
7 See 9 (4)
9, 7 German philosopher appearing from a milkman tune? (8, 4)
10 Slip up amongst the coconuts for a drink (6)
11 Hits the fishing vessels (6)
13 Remoulded brick has salty quality (8)
14 Old gin bottle broken opening veins (12)
17 Researcher of vintage riots (12)
20 Seasonal song in a state (North and South) (8)
21 Lemur of the kind rising after loss of monarch (6)
22 Medicine loses noble gas for Italian dynasty (6)
23 Hormone old cow not icy with (8)
25 Gallery of top Turners and Tracey Emin? (4)
26 Control of law (10)

Down

2 Misdirected mail comes for Daisy on taking brimstone (8)
3 Vera takes vanadium for an age (3)
4 Naming words sisters embrace love with? (5)
5 Fish found in narwhal I butcher (7)
6 Conductor of cations in motion? (9)
7 A kirk geared to Danish philosopher? (11)
8 German chemist rents nitrogen mixture (6)
12 Mix it with cool cider for something blue? (11)
15 Voila! Rise French chemist! (But a little difficult after the guillotine) (9)
16 Oil topic of debate for this person? (8)
18 Bears go wild for nuclear chemist (7)
19 Angle a lead compound makes? (6)
21 Dye I'll mix for happy scene after loss of energy (5)
24 Not home for last shout (3)

Crossword 5

Across

1, 8 down Restyle Ray's horror kit for Nobel Prize winner (3,5,5)
6 Ask Al a question and he's in a State (6)
9 Safari is off and some distance away (4)
10 Airlift ton? Straining! (10)
11 Neon tube compound (6)
12 Top lake (8)
14 Not the elbows affected by tennis joke! (4, 6)
16 Tim endures a spell in the joint at the beginning (4)
18 Ring leading 25 (4)
19 Pentacle so pearly (10)
21 Lead Jekyll's alter ego astray to compound (8)
23, 28 Char cleans or transforms environmental pioneer (6, 6)
25 Saint haters? No just not that keen on salt! (10)
27 American on the pull? (4)
28 See 23 across (6)
29 Electric exercises? (8)

Down

2 Get disease when you add nickel to a zinc flue (9)
3 Animal sounding husky (5)
4 Golly! Fox 'ere for fancy footwork! (11)
5 Mixed-up style in this ex-President (7)
6 Noble gas with halogen in atmosphere (3)
7 Worker at North Pole? No, exact opposite! (9)
8 See 1 across (5)
13 True praises for Louis' eponymous invention (11)
15 Euros gone on gratification (9)
17 Legume for primate fanatic (6,3)
20 Lab mice lost in old apparatus (7)
22 Camelid found in all amateur performances (5)
24 Rodent caught by decoy puppy (5)
26 Bird appears when tungsten taken (3)

Crossword 6

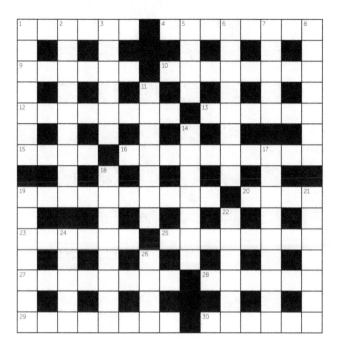

Across

1 First alkane loses head for next homologue (6)
4, 2 down Harpist torn for his smelling salts (6, 2, 9)
9 Something continental in amateur operatic production (6)
10 Spin alum ring for surface chemist (8)
12 Newer ads replied to after editing (8)
13 Tame baboon not neat when lost in the long grass (6)
15 Asked how this boat hidden away? (4)
16 Rocks found on Isle I hop to (10)
19 Rifle bottle (10)
20 Little sparrow appears if Pa confused (4)
23 Great untruth about this German chemist (6)
25 My French fowl caught by this animal? (8)
27 Rob stoic about automation (8)
28 Rat left theatrics for moral reasons (6)
29 Beryl makes rum cocktail bear fruit (8)
30 Something to capture the moment with a cream confection? (6)

Down

1 Mare led to green (7)
2 See 4 across (9)
3 Sodium ripe for this Scottish polymath (6)
5 Pasternak rants away on mountain top (4)
6 French car adds force for chemist (8)
7 Digit caught in that humbug jar (5)
8 Our ref's confused but made of iron! (7)
11 Test MEP in a storm? (7)
14 Encyclopaedist caught in candid erotica (7)
17 Libretto I redrafted for old fossil (9)
18 See 22 down (8)
19 Hunter and the hunted in their element (7)
21 Carefree Sian carrying flower (7)
22, 18 Eighth aria cast crime writer in new light! (6, 8)
24 Gerbil loses capital for capital (5)
26 Race to a tree (4)

Crossword 7

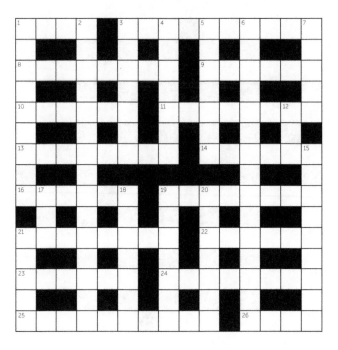

Across

1 Great story in the picture (4)
3 Crashes new fenestral detergent (10)
8 End union with such insinuation? (8)
9 Battered balalaikas alas removed from lake (6)
10 Line-up confused for Mr Wolf? (6)
11 Borax due for a change as a fungicidal mixture (8)
13 Zest we're mixing for small pincers (8)
14 Extract oil from ophiolite for green rock (6)
16, 7 down Blue oxide Hilda mixed without aid for structure unravelled by 23 & 24 (6, 5)
19 Part of a set diet? (2, 6)
21 Clove nun prepared not split (8)
22 Ivan weaves copper into cloth (6)
23, 24 Cowards can knit pair who elucidated 16 & 7 down (6, 3, 5)
25 Unlike dismal Iris to change (10)
26 Something on the beach by writer George? (4)

Down

1 Change diet ale to pale? (9)
2 Transform subarctic silver into people like you (15)
3 Rinse cerium away: honest! (7)
4 Mob rush to shape (7)
5 Stop barge overloaded with molybdenum (7)
6 Turn heroic shrew into much larger beast (5, 10)
7 See 16 across (5)
12 Treated thus on return on removal of sulphur (3)
15 Branded ark remade (9)
17 Have change now (3)
18 None move to poison without oxygen (7)
19 Dally with New Testament for physicist (7)
20 Drive id mad for separator (7)
21 Single outcome in gun we deployed (5)

Crossword 8

Across

1 Painful muscles magically transformed on taking aluminium (7)
5 Mixed bunch about German chemist (7)
9 Croatian ran out to animal (5)
10 Scattered sucrose about for supplies (9)
11 Silage can turn into painkiller (9)
12 Cloth recovered from brat we educated (5)
13 Puzzled pullet (e.g.) pulls leg to find relief (3-2)
15 Confused cadets rue enzyme (9)
18 Ropier few regroup for salvo (9)
19 Jelly has pickle inside (5)
21 Topographical feature in arid geography (5)
23 Maid all mixed up but in her element (9)
25 Tom lost on island with sight defect (9)
26 Fidelity in foxtrot Hermione performed (5)
27 Rare earth emits beryllium for another one? (7)
28 An ulcer connected with atomic theory? (7)

Down

1, 18 Alchemy afraid a scientist will emerge from the mix (7, 7)
2 Blast area for gypsum (9)
3 Gig about turned for Norwegian composer (5)
4 Crew rocks for sommelier's aid (9)
5 Rudimentary and alkaline (5)
6 Crabs and shrimps cat found in saucer (9)
7 Mother-of-pearl materialised when nitrogen added to measure of land (5)
8 What's left behind in 5 across contraption (7)
14 Spirit and Italian lager combined for sausage (9)
16 Old uranium transformed into another metal on loss of oxygen (9)
17 Ample bismuth and holmium in this mineral? (9)
18 See 1 down (7)
20 Oil stored in rustic amphora (7)
22 Left deuterium in the earthenware (5)
23 Prime Minister is right to change form (5)
24 Platonist philosopher loses us in loft (5)

Crossword 9

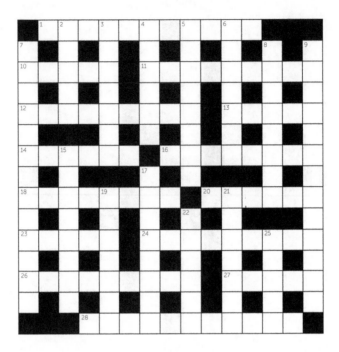

Across

1 Pertaining to union of Milo and Martina (11)
10 Contribution from a confused Russian president (5)
11 Lofty post for little one (9)
12 Radium counter malfunctions for story-teller (9)
13 German food in a German city (5)
14, 4 down A glum harvest composed (6, 6)
16 Crazy cat's entry into diary gets female support (8)
18 Duel with yeti at Christmastime (8)
20 Substitute a halogen with oxygen in formula for alloy (6)
23, 24 Cherish, envy and confuse chemist (5, 9)
26 Cubic loom weaves force of attraction (9)
27 Overweight in robes ermine (5)
28 Servant used by thrill seeker (11)

Down

2 Lavender hidden in Andrea's picture (5)
3 Domed building found around end of street (7)
4 See 14 across (6)
5 African grain that is supplemented by nitrogen (8)
6 Chemically transformed to black art on loss of two halogens (7)
7 Price highly so confuse the writing? (13)
8 Latest of assorted voices (8)
9 Great relation, if a little decayed (13)
15 Menus lie abot this element (8)
17 Reducible, reduced and extractable (8)
19 Your old fashion house returns for Adam's apple (7)
21 12 takes tellurium for bitterness (7)
22 A very old chap eats the bird as starter (6)
25 Thoughts about Sadie (5)

Crossword 10

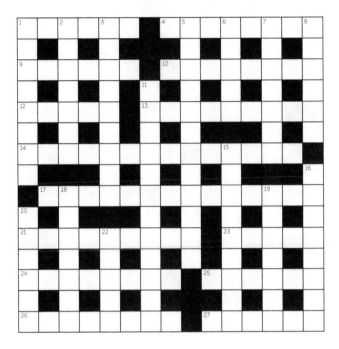

Across

1 Partition puts me in a spin (6)
4 Got acres of this slow food? Mais oui! (8)
9 Sir Cecil's island? Nominatively speaking! (6)
10 Artist has number of dynamic significance (8)
12 Somniferous ungulate (5)
13 Sons age me about toeing party line (2-7)
14 Mixed nitrites enable physicist (6, 8)
17 Unused zinc boot lost underground (10, 4)
21 Free from puce latex contraption (9)
23 Toxic mix of argon, sulfur and indium (5)
24 French aunt adds styptic to metal (8)
25 Don't fuse aluminium for English chemist (6)
26 Refuser who sounds like a horse? (8)
27 Confusingly go to seed for mineral curiosities (6)

Down

1, 20 Chaotic war in cessation with this polymath (3, 5, 6)
2 Rover in lead for saying (7)
3 A French ex swallowed pride, staying alive (9)
5 Intense games played by such things? (5, 7)
6 Girl sang about sweetheart (5)
7 Italian physician and physicist of prodigal vanity (7)
8 Pachyderm appears after street musician replaces first banjo with fourth guitar (6)
11 Lecture man on IUPAC definitions? (12)
15 Salt sounding like it would leave your taste buds in a pickle? (9)
16 Cut top off Italian mountains to reveal backbone of England (8)
18 Strange nanny marries copper (7)
19 Danish chemist lost in the Boers' tedium (7)
20 See 1 down (6)
22 Animal found returning to a mall (5)

Crossword 11

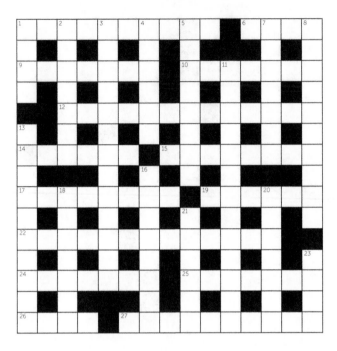

Across

1 Bold mohair creation of a certain shape (10)
6 Fish for learner mixing mother's ruin (4)
9 Roll with spirit of a revolutionary? (7)
10 Modified ark's got competitors in a speed trial! (2-5)
12 Aunt confused about treasurer: someone who might satisfy 8 down (13)
14 Half-cut, Lulu spins yarn about honesty (6)
15 French duck Ian follows taking right to North American (8)
17 Sounds like double-crossing cats (8)
19 Old music stirs Davy to discover this element, on loss of chlorine! (6)
22 Cosily chat with cop about bacteria (13)
24 Canine canter (7)
25 Confused animal takes extreme umbrage at such a compound (7)
26 Saint includes aluminium in seasoning (4)
27 Stem erupts from such woodland remnants (4, 6)

Down

1 Bewildered bore has habit (4)
2 Muddled moron in charge of letter (7)
3 Crazy crab cracks the long-handled instrument (13)
4 Something very cold in juice captain drank (6)
5 The French regain this North African (8)
7 Victorian PM loses head to Middle Eastern person (7)
8 Matron goes wild about epicure (10)
11 Where justice is meted out to marsupials? (8, 5)
13 Logic's used to create such compounds? (10)
16 Halt core meltdown with coordination compound (8)
18 Leg knave broke on loss of king? Good news (7)
20 Metal drum I thrice played (7)
21 Can't include in Old English compound (6)
23 Spoils planet (4)

Crossword 12

Across

7 Bertha gets letter on taking rhodium (4)
8 Mercurial idea transforms into chemist after lad leaves (5, 5)
10 Tests out of order for fattest? (8)
11 Keep lawrencium mixed for German astronomer (6)
12 Mohr works bismuth into shapes (6)
13 Element of the continent? (9)
15 Dubious ref twists leg, resulting in high score? (6, 7)
18 Mutate or change: the same! (8)
20 Add cobalt to 26 across for girl (6)
22 Sulfur meant to turn into flower part? (6)
24 Shaking large tin essential! (8)
25 Exhort mice to create heat? (10)
26 River taking a meandering line? (4)

Down

1 Aped about hearth for polygons (10)
2 Cleaner abhorrent to nature? (6)
3 Royal imp takes the French air (8)
4 Bake erbium in this glass? (6)
5 Perpetrator confused clue for botanist (8)
6 Oddly laid my egg for fruit? (4)
9 Grind their durum for dubnium (13)
14 Soluble mix of vanadium and sodium? Impossible! (10)
16 Lutetium metal I crafted into best (8)
17 Gets nice about science of heredity (8)
19 Russian chemist ditches Eve to become a botanist, pioneer of 17 across (6)
21 Convincing gent follows company (6)
23 Titanium axe loses edge, recast in cab (4)

Crossword 13

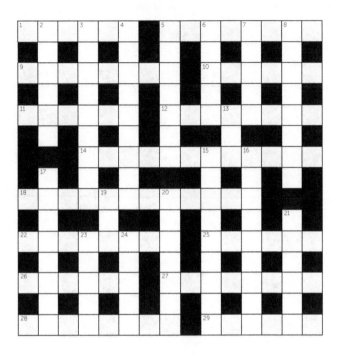

Across

1 Ma returns to French father and physicist appears (6)
5 Ha! Look-am in a state! (8)
9 Philosopher put larch about (8)
10 Gals, that is, turn to cattle fodder (6)
11 Token couple on loss of the Spanish (6)
12 Lime tutu has element of Paris about it (8)
14 Recast arty amphorae for alternative medicine (12)
18 Split sachets aim to provide styptics (12)
22 Ox in area creates disorder (8)
25 Selenium identical to this seed? (6)
26 I mix stain for this indole derivative (6)
27 Poet is so confused by similar atoms (8)
28 Mathematician upsets large nag (8)
29 Must turns to wine on addition of calcium (6)

Down

2, 17 Recast Mike as flamingo for relief! (4, 2, 8)
3 Confused camels opt for outer layer (9)
4 Sourer one all at sea and wrong (9)
5 Cop he liaised with included daughter of Polonius (7)
6 Sounds like lean pianist (5)
7 Even the owl is stunned by this composer (5)
8 Chiefs mix tungsten and uranium with gum, ending Members of Parliament! (8)
13 Oddly envied girl (3)
15 Confused morals hit by such a tempest? (9)
16 Obstructive circuit components? (9)
17 See 2 down (8)
19 Decapitated hero turns to rock (3)
20 Neil in a muddle for amino acid (7)
21 Protozoan warms one's beau evenly! (6)
23 Root around bottom of drawer for engine part (5)
24 Non-exclusive element returns to begin with (5)

Crossword 14

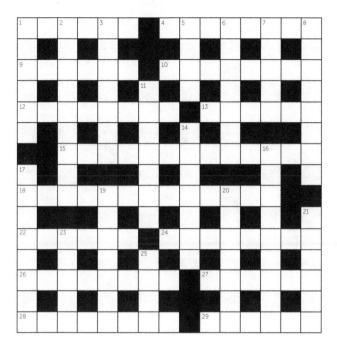

Across

1 I came to find cerium in the Floating City (6)

4 Perfectionist lets Rick adjust (8)

9 See 5 down (6)

10 Sage turns west for something lacking brimstone (5, 3)

12 Chose our mixed colour (8)

13 Last radium compound found in the stars? (6)

15 0.000001 (3, 2, 1, 7)

18 Thatcher cross about psychological assessment (9, 4)

22 Chemist involved with many holmium compounds (6)

24 Prose pro composed for The Tempest? (8)

26 Swift creation of a nation (8)

27 Release from Ganymede's orbit (6)

28, 2 down Our ox breed nothing to person close by (4-4, 9)

29 Need indium compound for polycyclic (6)

Down

1, 25 Ought vanadium, carbon or iodine transmute into this French playwright? (6, 4)

2 See 28 across (9)

3 Care about Ben in compound (7)

5, 9 across Cretins row about bellmen (4, 6)

6 Confused loser takes cobalt for compound (7)

7 Returned regal drink (5)

8, 17 No-frills kid ran an awfully unacknowledged contributor to elucidation of structure of 21, 23 down (8, 8)

11 Former French name confused in quad (7)

14 Camphor also inside pot (7)

16 Zenith one changed for peak performance (2, 3, 4)

17 See 8 down (8)

19 Collie dog loses tail and gets dirty (7)

20 Step around a German sculptor (7)

21, 23 Two screws holding the key to life? (6, 5)

23 See 21 down (5)

25 See 1 down (4)

Crossword 15

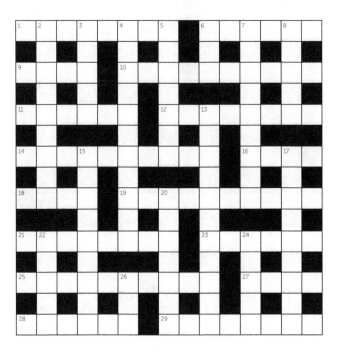

Across

1 Copper coils around end of spring: sweet! (8)
6 See 5 down (6)
9 Egg on very unfit man initially (4)
10 Anytime has muddled muscular weakness (10)
11 Add sulfur and selenium to raw fish (6)
12 The French country house of a mathematician? (8)
14 Lettuce Ross found in loo: laugh out loud! (5, 5)
16 I hurried to this country (4)
18 Val's return to Eastern European (4)
19 Can agree an arrangement of this polysaccharide (10)
21 Caribbean islands worker unwell with extreme endometritis (8)
23 Name on the counter? (6)
25 Idiom I'm sure about loses energy for metal (10)
27 Rancid removal of radon not basic! (4)
28 The old Rolf returns for famous pharmacologist (6)
29 Mistakenly takes wig: most awkward! (8)

Down

2 Lip lover finds nothing in Northern city (9)
3 Philosopher lost in harmonica music (5)
4 Misconstrued mystic realm perfectly balanced? (11)
5, 6 across Hardliners caw about evolutionist (7, 6)
6 Did not lose charged particle for insecticide (3)
7 Look at meanie Rex again (9)
8 Punk left unpicking top of cake (5)
13 Make mega ego grow into physicist (6, 5)
15 Play the French ivories for chemist (9)
17 Mecca aids those of an intellectual disposition (9)
20 Trust in galvanization for the corrosion within (7)
22 Sodium salt getting up your nose? (5)
24 Oddly I'd read quiz to Middle Eastern person (5)
26 A lively Ella took leaf (3)

Crossword 16

Across

1, 6 down Limiting war while confusing the digitalis doctor (7, 9)
5 Was kind to evolutionary biologist (7)
9 Lead other inside for plant (3,4)
10 Sounds like a pamphlet to appeal (7)
11 Nun's verbosities about something unimposing (15)
12, 20 down Ore in gorge well mixed for English author (6, 6)
14 Limit ray to army use? (8)
17 The French crying over alcohol (8)
18 Silver raid bungled in Morocco (6)
21 It's sad orienting such breakups (15)
24 Stair Al returned to with ropes (7)
25 Serious bear nestling inside (7)
26 Rep in spin about little ones (7)
27 Old graphite loses edge – what a pain! (7)

Down

1 Massive district of London? (7)
2 Lily got holmium confused in her study of rocks (9)
3 Still amongst finer things (5)
4, 22 Robust imp is twisted (6, 5)
5 Confused comic I'd aid becomes doubly corrosive with molybdenum deficiency (8)
6 See 1 across (9)
7 Wandering wise man has energy for picture (5)
8 Say it's fluorine content (7)
13 Fruit hidden by ogre engagement (9)
15 Dared moan about constellation (9)
16 Asses die of such ailments? (8)
17 Finnish chemist loading compound (7)
19 Otto is right about this dish! (7)
20 See 12 across (6)
22 See 4 down (5)
23 Buzz with smooth rum inside! (5)

Crossword 17

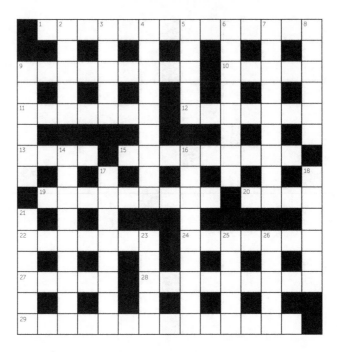

Across

1 Cosmologist thinks when page is turned (7, 7)
9 Ruse Degas employed for removal of interference (9)
10 See 18 down (5)
11 Refer it to metallic compound (7)
12 Peg's old spot (7)
13 Prompts 1-29 (for example) without final call (4)
15 Rural mice scatter like quicksilver (9)
19 Author mixed up in game. Why? (9)
20 Even Nigel has a diagram (4)
22 Extremely heated teach tormented, plotted and arrived (7)
24 Body parts in salve Oliver applied (7)
27 Dupe takes ringleader for small crustacean (5)
28 Language of no repeats, confusingly! (9)
29 Dressed like an Amazon? (6-8)

Down

2 Feisty German hides flammable feline (according to one poet) (5)
3 Gold applied to lip for exclusion principle? (5)
4 Segregate 22 in April (6, 3)
5 Ruth's upset by injuries (5)
6 Extreme foul! Ref was confused by prattlers (8)
7 Brighten tired aria (9)
8 Norwegian composer confusingly takes first exit for German physicist (6)
9 Spoil café Ed restyled (6)
14 Yes, retina upset by this condition! (9)
16 Fruit with a sign of cancer? (4, 5)
17 Compound hit by plane when T.A. left (8)
18, 10 across Fierce minor turns to Italian physicist (6, 5)
21 Outlined in rash ape suffered (6)
23 Be wed to fool? (5)
25 Take out virtuous agent (5)
26 Cat inside pounces (5)

Crossword 18

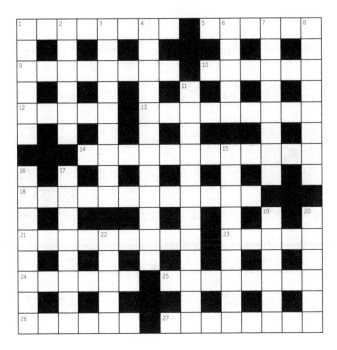

Across

1, 25 A saucier go-go girl changes for father of mineralogy (8, 8)
5 Refreshment novice teacher provided (3, 3)
9 Small unit hidden in Wolfgang's trompe l'oeil (8)
10 Positive beaut crazy about phosphorus (6)
12 I nose around compound (5)
13 Something essential Lionel lost in Rio (6, 3)
14 Oxidising agent Panama Regent prepared (12)
18 Philosopher transforms sentient twig (12)
21 Poem Ellen composed for cooking ingredient (5, 4)
23 Stadium Arsenal lose top soccer league in (5)
24 Reggie upset physicist (6)
25 See 1 across (8)
26 Within the best ancient city (6)
27 Battery I turn into rare earth oxide (8)

Down

1 Banger explodes in rift valley (6)
2 Organism with nickel deficiency in ecstasy? (6)
3 Bergen tug scrapped for printer (9)
4 Muddled mundane tenor not showing off (12)
6 Oddly crumpled linen hides goldsmith's vessel (5)
7 Tale about tote-no booze allowed! (8)
8 Stella in confusion about West Indies (8)
11 Discover out of the darkness (5, 2, 5)
15 In Zaire, Al becomes madder! (9)
16 Dusk with gilt fusion (8)
17 Main poet transforms putrefying compound (8)
19 Let go when bored about brimstone (6)
20 Sarah makes a change in desert (6)
22 Relative indigencies even (5)

Crossword 19

Across

1   Jade's upset about Sir: he invented the Pill! (8)
6   Medicine for wise man after party? (6)
9   Seemingly coloured compound within (6)
10, 2 down Loiter around fjord ice for French radiochemist (8, 6)
11 Organic chemist has word about woad (8)
12 Zirconium ash I turned red (6)
13 Lose obscene comic initially for redundancy (12)
16 Aviators who don't know the words? (12)
19 Imam maligning animal inside (6)
21 Confusingly use dinar as coinage (8)
23 Neutered merino extremely particular (8)
24 Elemental Norse god that is surrounded (6)
25 Lose hour from mixed harvest: go hungry! (6)
26 Metal men churned to cheese (8)

Down

2   See 10 across (6)
3   Ran account into red? (5)
4   Vassal ran around magic bullet (9)
5   Unbeliever astray in field (7)
6   Padre's sermon involves raiment (5)
7   Spices men mix for samples (9)
8   Dunce a GI bothered for direction (8)
13 More tomes about pressure gauge (9)
14 Sir confuses mutton for metal (9)
15 Transmute beryllium and palladium dust-capital!(8)
17 Bores Ed about ulceration (7)
18 Country of money. Literally (6)
20 Role 24 plays in French river (5)
22 Find gas with road turning North (5)

Crossword 20

Across

1 Thumped dunce got pH confused (7)
5 See 3 down (7)
9 Bring it around to restrain horse (7)
10 Muddled multiethnic etc removed for element (7)
11 Russian chemist mistakenly delivered me mint (6, 9)
12 See 18 across (6)
14 Cup leper prepared for herbalist (8)
17 Restrain wild areas (8)
18, 12 across Others trying to unravel this idea? (6, 6)
21 Confused choristers elope for analysis (15)
24 Path a Northerner took to find hidden furnace (7)
25 Internal mail a mosque returned to African country (7)
26 President in style confusion (7)
27 Any salt compound for chemist (7)

Down

1 Tried mixing 5 down for olivine (7)
2 Ran online puzzle-not straightforward! (9)
3, 5 across Preachers sob about ammonia production (5, 7)
4 Smut a leading Dame returned for points (6)
5 Element symbolic of an Italian river? (8)
6 Ten photos developed there and then (2, 3, 4)
7 Even he'll hit the top! (5)
8 Mar a small sovereign for tea with 11 across? (7)
13 Treats can turn into chemicals (9)
15 Dissenting chemist sounds clerical (9)
16 Radon ran amok for European (8)
17 See 19 down (7)
19, 17 Get that parsley prepared for treatment (7, 7)
20 Blended hashish not hot in this pipe! (6)
22 Radical lye and thorium compound (5)
23 Strum balalaika for concealed dance (5)

Crossword 21

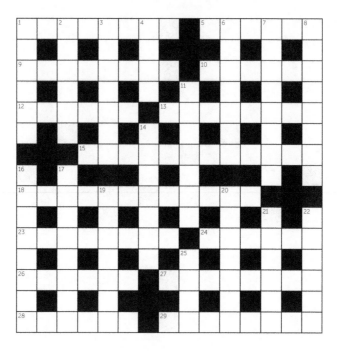

Across

1 Shark ova hatched by physicist (8)
5 Bored about brimstone? Let go! (6)
9 Mound due to small intestine? (8)
10 Bad oar lost on boat (6)
12 Togetherness when last Tories join union! (6)
13 Sir Humphry's light for minors? Sounds like it (4, 4)
15 Senses clerk's confusion about imprudence (12)
18 Chemist Fritz takes festive reindeers for ribbon sellers (12)
23 Goon acts weird: figures! (8)
24 Extremely alarming element symbolically (6)
26 Violet in her element! (6)
27 Scales up tablets (8)
28 Tries to write dissertations (6)
29 Hot, the French men turn to alcohol! (8)

Down

1, 16 Clan scatters this salt symbolically (6, 8)
2 Aluminium OK in moulded clay? (6)
3 Face toned by concealed chemical (7)
4 Egg constructed of molybenum, vanadium and uranium (4)
6 My robes altered for very little babies (7)
7 Cox Al ate slowly contained cause of kidney stones (8)
8 Bad set-up in capital (8)
11 Each cut transformed by this plant extract (7)
14 Bread basket found in souk rained upon (7)
16 See 1 down (8)
17 Debit Seb wrongly for natural phenomena (3-5)
19 Reputable agency loses ale at pub. Period. (7)
20 Scare in school involves test for arsenic (7)
21 Aluminium in river of Arthurian legend (6)
22 Loser returns to top chemist for compound (6)
25 For example, Actinium returned in this structure (4)

Crossword 22

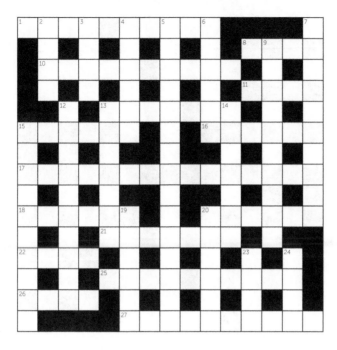

Across

1 Press papaw medley for vespine varmints (5, 5)

8 Computer language of Indonesian origin? (4)

10 In Japan creating juice that aids digestion (10)

11 Radical lacy creation (4)

13 Staff in mixed company leave (4-3)

15 Crown Prosecution Service present chemist turned novelist (deceased) (1, 1, 4)

16 Sorted Sicilians without sin in compound (6)

17 Pedestrians' tans surprisingly lift mood! (15)

18 Pork pie large for this German chemist (6)

20 See 19 down (6)

21 Titan prepares odd brew for Asian (7)

22 Wrestler in sour mood oddly (4)

25 Miner made it big with this rock (10)

26 Rod upset Alex (4)

27 Optimist's glasses do interest, confusingly (4-6)

Down

2 Pals astray in mountains... (4)

3 ...and French sodium found in volcano (4)

4 Was raw about capital (6)

5 A vainer huntress dissociating chemist (6, 9)

6 Fortissimo kiss confused for smacks (6)

7 Straw bound for friendly donkeys by the sound of it! (10)

9 Notice urea compound for vendor (10)

12 Hermit also same temperature on mixing (10)

13 Channel ion into duct (7)

14 Fashionista loses a hat: confused showing cleavage (7)

15 GB's first in line's rule? Speaks volumes! (7, 3)

19, 20 Confuse singer with grog re Fred's dancing partner? (6, 6)

20 Loquacious mammal? (6)

23 French end first night for Scandinavian (4)

24 Practice passed on in some meadow (4)

Crossword 23

Across

1 Lied about result for metalloid compounds (10)
7 Oddly cry as it forms a sac (4)
9 Artist remodels her bulge (8)
10 Gypsy Mum lost my plaster (6)
11 Beetle cars a top bloke redesigned (6)
13 Brag about harp on histogram (3, 5)
14 Join quasi compound-see a certain something (2, 2, 4, 4)
17 Geeks apply phenol to itches (12)
20 Romantic retreat with nothing in it? (4, 4)
21 I confuse Mum's love for something elemental (6)
22 Titanium a hit within island (6)
23 Terrapin fin-it yields an endless quality within (8)
25 Even earth or mud contains this particle (4)
26, 3 down Troubled tribesmen leer at web creator (3, 7, 3)

Down

2 Fear a chest infection hides other complaints (8)
3 See 26 across (3)
4 Haber confused by abstention (5)
5 Composer lies in crumpled bed (7)
6 Polyhedral sucrose (5, 4)
7 Confused Curie in comp for radioactive element (11)
8 Tree remains on cricket pitch (6)
12 Inert gum one mixes for synthetic element (11)
15 Transmuter Michael lost in street briefly (9)
16 See 24 down (8)
18 Arm unit to provide Davy's discovery (7)
19 Calcium blot transmutes to other metal (6)
21 Cabinet leader leaves office for liquor store (5)
24, 16 Question Royal Navy about uranium? Doesn't follow! (3, 8)

Crossword 24

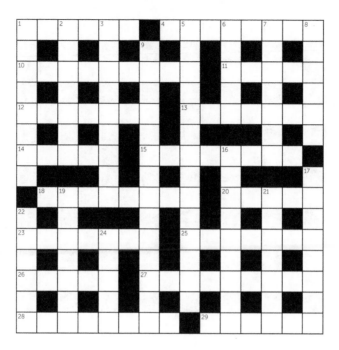

Across

1 Pal wed, returning for wattle (6)
4, 12, 22 down Engineer bard's ode about liking numb arm (8, 7, 6)
10 Victor was William! (9)
11 Deb yearns to start with chemist (5)
12 See 4 across (7)
13 Camel I'd confuse with numbering system (7)
14 No tin around this element (5)
15 Last performance of a mute perhaps? (8)
18 The elder confused the unprepared king (8)
20 Being pulled in Acton Town (2, 3)
23 Hair gel concocted for Elizabethan explorer (7)
25 Element of lost innocence initially (7)
26 Diana upset by nymph (5)
27 Man using uranium compound? Not funny! (9)
28 Large nag lost by French mathematician (8)
29 Genus I introduced for gifted one (6)

Down

1 Cleaning 1920s art supremo? (8)
2 Ironmongery beloved by barnstormimg fanatic? (4, 3)
3 Plenty of dance after a bun? (9)
5 Disregard law re confused chemist-composer (3, 6, 5)
6 Doctor in some dictionary (5)
7 White Album engages at beginning (7)
8 Lives in Llandrindod Wells (6)
9 Upsettingly rubs arm's length for such radiation (14)
16 Ref shouts about quick temper (5, 4)
17 Methane from the mire (5, 3)
19 Conversation taking liberties initially (7)
21 Bits of lithium iodide around capital (7)
22 See 4 across (6)
24 Continent hidden in diagram (5)

Crossword 25

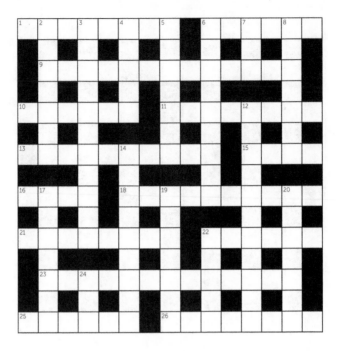

Across

1 So heroic about such a curve (8)
6 I'm confused about 16 across in Picture House (6)
9 Lead poisoning turns to cleric's pain (8, 5)
10 Make a meal out of superphosphate (potash he rejected) (6)
11 Crushed granites emit radon! (5, 3)
13 But her corn mix provides colour (5, 5)
15 Tough lady had to lose: not pretty! (4)
16 Cancer treatment with chromium deficiency leads to skin complaint (4)
18 Real forces form such sources of carbonate! (5, 5)
21 Such slumber repels me awfully (3, 5)
22 Try a rhenium compound for this part of the anatomy (6)
23 Enliven bandit about window shade (8, 5)
25 Alex I adore with such blindness to begin with (6)
26 Randy ox turns to stone with brimstone! (8)

Down

2 Pukes up over suicide (7)
3 Can she zip 'em misbehaving apes? (11)
4 Open University term begins with target-setting (5)
5 Abandoning top doctor, her child turns to pioneering chemotherapist (7)
6 Inventor who sounds off loud in the morning? (9)
7 Liner loses her majesty, returning with nothing (3)
8, 20 Camel afraid about hay turning into scientist? (7, 7)
12 Routine aunt performed for small particle (3, 8)
14 Senior chest rattles with something musical inside! (9)
17 Teacher vilifies herb within (7)
19 Criticise sayings on carpentry tools? (7)
20 See 8 down (7)
22 Bream turns the colour of caution (5)
24 Pollutants noxious to begin with! (3)

Crossword 26

Across

6, 23 Eponymous equation for girl on rheumatism treatment (8, 8)
8 August chemical dreamer (6)
10 Capital Anne returns to after six (6)
11, 7 down Lap unmerited due to dense metal (8, 7)
12 Wrongly equip shy bodies (9)
13 1 down transformed and relocated to South America (4)
15 Under calcium oxide: glorious! (7)
17 Thallium coin forged for President (7)
20 Garden herb costing a fortune? (4)
21 Pa tumbles over salts (9)
23 See 6 across (8)
25 Went on scramble for physicist (6)
27, 5 down Locate bulb variety of elemental colour (6, 4)
28 Animal trailing around (8)

Down

1 Mail gone astray in Africa (4)
2 Reg leaves regency silver for bureau (6)
3 Drum I played on three occasions for heavy metal (7)
4 Animals in book a pisky read (6)
5 See 27 across (4)
7 See 11 across (7)
9 Merlin mixes potassium in Red Square (7)
12 Scientist excluded on principle? (5)
14 Particles of uranium in Belgian battlefield (5)
16 Capital of elemental significance (7)
18 Ban Leon from wandering around this country (7)
19 Seasonal sounding synopsis (7)
21 Cornucopian quanitity (6)
22 Sounds like blood and tears surround this chromatography pioneer (6)
24 Brainbox under corporate analysis? (4)
26 Provident poet within (4)

Crossword 27

Across

1 Dies if shakiness leads to this ailment (8, 7)
9 Famished Mennonites within (5)
10 Tall rogue turns into drunken yob (5, 4)
11, 5 down Campanologist's conditioned canine! (7, 3)
12 Chirpy about first romantic dance (7)
13 Cart lithium around for perfume ingredient (6)
14 Thing about the evening (5)
17 War-worn age oddly incorrect (5)
18 Ejected from seat of power? Sounds like it! (6)
21 Dream about silver leading to radiation overdose (7)
22 Toll I am returning for tights (7)
23 Bird dregs colt digested (9)
25 Cat he trained to instruct (5)
26 Shaver saturnine about Swedish chemist (6, 9)

Down

1 Chief rapper cool inside! (6)
2 Sky package Viv created for drudge (6)
3 Chats about recordings of physicist's feline (12, 3)
4 Pictorial art duellist transformed (11)
5 See 11 across (3)
6 Propulsion pioneer wrongly trifles with rank (3, 5, 7)
7 He got map mixed up for saying (8)
8 Silently ate toxic compound (2, 6)
12 Instrument for measuring subcultural slang? No, optical activity! (11)
15 Swaps mag for methane (5, 3)
16 Summon Goliath to hidden country (8)
19 Base aluminium destroyer (6)
20 Draws corrosively (6)
24 Are about age (3)

Crossword 28

Across

1 Bung chap exercise kit (5-3)
6 Uncover Edgar Allan in sex romp (6)
9 Carolers go around successful footballer (10)
10 Addict takes first train leaving from Ulster (4)
11 Hoodwink Wiggins for shrink (5, 7)
14 Suddenly came out of wasp range (6)
15 Notorious americium fusion (8)
17 Documented re condor confused (2, 6)
19 Off radon after acid treatment (6)
20 It's science I'd employ for such killers (12)
23 Odd chop meal creates unconscious state (4)
24 He then pans around compounds (10)
26, **16 down** Chemist reminded vet about lime (6, 9)
27 Turnover voter at penultimate ballot (8)

Down

2 Strange appearances from buffoons even! (4)
3 Larch done over with example of 20 across (9)
4 Auntie crabby about Ray leaving (3)
5 Gecko taking iridium treatment composed (7)
6 Physicist nicer if more confused (6, 5)
7 Gold LP I played for physicist (5)
8 Sue chemist about something 5 down might write (5, 5)
12 Experts confused sect in Congo (11)
13 From posing to a type of encephalopathy (10)
16 See 26 across (9)
18 Queen pined about river (7)
21 Clever sting (5)
22 Paws fester badly with strontium deficiency (4)
25 Bird tight after taking quicksilver (3)

Crossword 29

Across

8 See 23 down (6)
9 Sprite extremely naughty with speed (5)
10 Eponymous cell lined with aluminium (7)
11 I mash us around for synthetic element (7)
12 Bogey men in dog rescue (5)
14, 15 across Air around gents is noble! (5, 3)
16, 28 across Owls emphatic about aeroplane (7, 5)
18 Composer deliberates without art out east (7)
20 Snoop odd party (3)
21 Tests Ma's ex confusingly (5)
23 Optical amplifier from real sulfur compound? (5)
24, 29 across Aria on television transforms French chemist (7, 9)
26 Devil cod in kiln? (3, 4)
28 See 16 across (5)
29 See 24 across (9)

Down

1 Group can include left (4)
2 Sect confuses extreme alphabet for old Mexicans (6)
3 Scattering her gaily about? (8)
4 Hones uranium compound to remove footwear (6)
5 Sounds like sugar for smart Alec! (8)
6 Impish group of warriors to begin with (4)
7, 8 Physician and writer based methods around oxygen (6, 7)
13 Give back a prey perplexed (5)
15 Gibbons no less for thermodynamicist (5)
17 Incredible lack of chromium uneatable (8)
18 Confused so devils disappear? (8)
19 Footballer failing to turn up to work? (7)
20 Pack odd lens for physicist (6)
22 Area around boil even is circular (6)
23, 8 across Waltzing with old numb physicist (6, 9)
25 Dimension confusingly limited after lifting lid (4)
27 Flower in the eye of the beholder? (4)

Crossword 30

Across

7 Blended booze rules at this temperature (8, 4)

9, 10 across Physicist enables nitrite compound (6, 8)

11 Pascal sure about this alchemist (10)

12 See 25 across (4)

13 Second lesson about scenic knowledge (7)

16 Bewildered alien in compound (7)

19 Lab uses leading mathematician to create animal (4)

21 Compound arcane about then (10)

24 Well-mannered conflict? (5, 3)

25, 12, 4 down Nanny defrosted roll for Poet Laureate (6, 4, 8)

26 Arc alignment altered for projector (5, 7)

Down

1 Deny Ben agate creation (8)

2 Courageous container? (6)

3 Eyes pup with good digestion (7)

4 See 25 across (8)

5 Radical confused plenty! (6)

6 Iron lost in post turns particular (8)

8 Insect a trifle agitated within (4)

14 As I march around, something alluring develops (8)

15 Eric had wicked physicist inside (8)

17 Matches ulcers if treated (8)

18 Confused earl ten for ever (7)

20 See 22 down (6)

22, 20 Irritable dog lost in Venice (6, 6)

23 Turnip even knee-deep (4)

Crossword 31

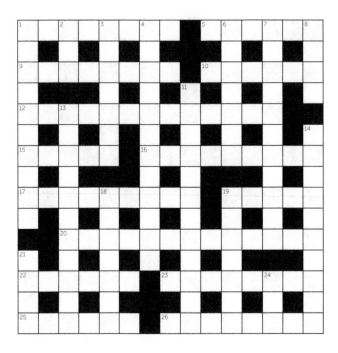

Across

1 Lacking sunlight vitamin, fresh food lost at sea (8)
5 Loves silver primates? (6)
9 Imp cries about experimenters (8)
10 Online warming effect? (2, 4)
12 Mix magical ochres for farmer's helpers (13)
15 Composer involved in showbiz ethos (5)
16 Confuse her mystic science (9)
17 Notary raw about Scandinavian rodent (6, 3)
19 Sounds like this holy man is in the chippie (5)
20 Polio persists strangely, revealing another ailment (13)
22 Snakes around Dad's outer ear (6)
23 Let orbit spin into lyrics (8)
25 Compound some tin particles (6)
26 Inspiration aping ape upsettingly (8)

Down

1 Abandon men to very individual orchestra (3-3, 4)
2 Dandy oddly floppy (3)
3 Chariot overturned and in a stew! (7)
4 Her carols can transform environmental pioneer (6, 6)
6 Physicist upsets vain gal (7)
7 Plants turned into pasties (11)
8 Forecasts now include precipitation (4)
11 Book by 4 down printing less, disturbingly (6, 6)
13 Rod lazes around bar with little shavers (5, 6)
14 Confused, Porgy chose this instrument (10)
18 Analgesic is right about pain (7)
19 Old language for translation to begin with (7)
21 Prime minister takes the French back and gains upper hand (4)
24 Odd tramp dance (3)

Crossword 32

Across

1 Ban bad positrons from these photon phenomena (10, 5)
9 Copper and zinc alloy that is about underwear (9)
10 Upsettingly nails gastropod (5)
11 Wrongly cane hen to improve (7)
12 Gigantic mother gypsy (for example) (7)
13 Odd trout just a small one (3)
14 Puck upsets brother at charcuterie (4, 7)
17 Satirical old hand initially takes no notice of things past (11)
18 Even Brenda has this natural acid (3)
20 Birds mad before start of summer(7)
23 Punt for leading position adopted by 9 across (2, 5)
25 See 24 down (5)
26 British weather Ian followed with East European (9)
27 Waltzing about numb old physicist (6, 9)

Down

1 Bit mean about this type type of music (7)
2 Pasta prepared past eight (9)
3 Gum tangled reins (5)
4 Upset her retort in the French Revolution (3, 6)
5 Fuming about mule taking oxygen? (5)
6 Blames tea for such an element (4, 5)
7 Gain overall lead for Kiwi girl (5)
8 Hurls plutonium, transmuting to non-metal (7)
14 Pour iridium into pot for sweet-smelling medley (9)
15 French physicist unusually queer about beryllium and chlorine (9)
16 Mohair can transform this instrument (9)
17 Harry cool about drink (7)
19 Nitrogen action about radon isotope (7)
21 Diced to power thrice! (5)
22 First question about bus I turned into explosive device (5)
24, 25 across German chemist transmutes zirconium bath with fire (5, 5)

Crossword 33

Across

1 Alchemists' small department in outskirts of Athens (6)
5, 9 Haven retrains us for Swedish chemist (6, 9)
9 See 5 across (9)
10 Gulp back stuff (4)
11 Tap ore on principle (6)
12 Mania about the blood molecule (8)
14 Sea dog (8)
16 Small particle in disturbed moat (4)
18 Lacking nitrogen, overturn futon for food (4)
19 Lousy ref for current cruciverballist! (8)
21 Not bad twelve months for vulcaniser! (8)
22 Chimp acting with internal influence (6)
24 Money-changing silver before odd idol (4)
26 Kitchener is gaudier with iodine and sulfur instead of neon (9)
27 Rome is mistaken for something else (6)
28 Basic tellurium in Scottish lake (2-4)

Down

2 Dad's odd about a foreman, even deceased! (4,2,1,4)
3 Pure brimstone hidden in money-bag (5)
4 Doe stirs around hormones (8)
5 Mess around extreme Tory for modus operandi (6)
6 Pray about Hal's indefinite article being particles (5,4)
7 Neutrino in even stratum (3)
8 Nobel laureate expresses horror about Ray's kit (3,5,5)
13, 23 Solicit reception about critical pH (11,5)
15 Circular gear for nematode? (9)
17 Lovelock's hypothesis about cobalt and lutetium compound (8)
20 Befuddled, break the last lab glassware (6)
23 See 13 down (5)
25 Odd grass ideal at times (3)

Crossword 34

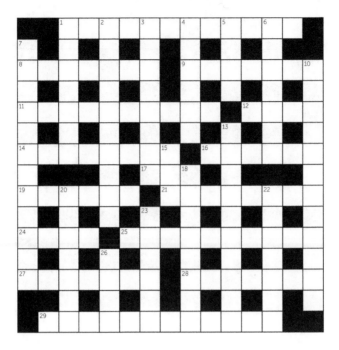

Across

1 Smash usual racquet in very wet farms (12)

8, 9 Tenor did complain about goo (7,7)

11, 12 Shy diva mixes up myrrh for Cornish chemist (3,7,4)

14 Hostilities upset mine site (8)

16 Chip's last party around old medicine (6)

17 Oddly rheumy type of sleep (3)

19 Lacking sulfur, Mum is so upset about this element (6)

21 Tact lays about reaction promoter (8)

24 Learner leaves novel concoction for cooker (4)

25 Confusing Earl's quest leads to compound fracture (10)

27 Last elementary metal, symbolically (7)

28 Windy Ian takes aloe mixture (7)

29 Opening comment about extremely crude men (12)

Down

1 Man Gran upsets like 13 down (7)

2 Quite dubious about Ed leaving common (10)

3 Cough that is right for casino worker (8)

4 Wrongly dials last number for remote sensing equipment (6)

5 Upset Mum about a damselfly (4)

6 Anne joins Des in groups of nine (7)

7 Upsettingly employ egoist for theory of knowledge (12)

10 Exactly about eye on gas (12)

13 Newest oath about bridge (10)

15 Speech oddly dry! (3)

18 Spot cute Lama confused (8)

20 Virtuoso plays a most extreme reggae (7)

22 Roy's kit mistaken for Plantagenet supporter (7)

23 Dutch physicist lost in neon maze (6)

26 Strict company (4)

Crossword 35

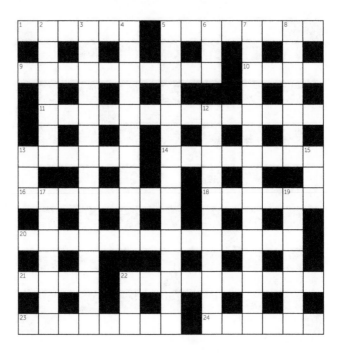

Across

1 Symbolically osmiridium is Egyptian deity (6)
5 Lowest weight sounds like polygraph (8)
9 Upsettingly decline pal plumbago (6, 4)
10 Old cow lost in ancient city for sauce (4)
11 Analyses song about nativities (14)
13 Scandinavian god I'm at odds with for oxide (6)
14 Spock crashes car into geological strata (8)
16 Devon town on the slate? (8)
18 Hell where busybody hides (6)
20 Nag rebel about league: bloom can result (4-5, 5)
21 Mass of potassium oil mixture? (4)
22 Headgear disturbing soft crania (10)
23 Ray heard about this Welsh town (8)
24 No, Eve's confused nonetheless (4, 2)

Down

2 Things we dish to hidden Scandinavians (7)
3 Record every memo about past recollection (9, 6)
4 Startles ref about go-getter (4-7)
5 Beginners docile about this apparatus (6, 9)
6 Geoid's odd deity (3)
7 Perform a post-mortem pirouette on returning vase! (4, 2, 4, 5)
8 Pulling article, ask Russian President about satellite (7)
12 Prepared ham pie meant for speed (11)
13 Twitch stick in the middle (3)
15 Odd Scots requiring assistance (3)
17 Restyle leg-hair of explorer (7)
19 Bores U2 about abounding (7)
22 Capacitance unit loses radium on whim (3)

Crossword 36

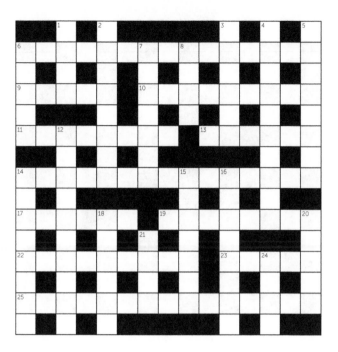

Across

6 Periodic chemist delivered me mint condition (6, 9)
9 Fundamentally caustic (5)
10 Mathematical regime around odd Croat (9)
11 Grime around Rio transforms into magic book (8)
13 Coil around odd face coverings (6)
14 Waltzing around dumb old physicist (6, 9)
17 I wait around last stop for elk (6)
19 Lend swan to chemist (8)
22 Crustaceans selfish about outer loch (9)
23 Apostle I follow on principle of exclusion (5)
25 Bar nervous dogma about molar constant (9, 6)

Down

1 9 across loses carbon compound for influence (4)
2 Bond producer of brassica? (8)
3 Clare confuses ecstasy for breakfast (6)
4 Her cleanup messed up by Irish creature (10)
5 Dispossession of confused IT novice (8)
6 Clean bewildered budgie without iodine (5)
7 Wrongly grab hem in Africa (7)
8 None around for noble (4)
12 Hop in olden compound (10)
14 In Wales I wandered in the Proterozoic (8)
15 Cheese left around bloodsuckers (7)
16 Seals plush azo compound (8)
18 Cloaks aluminium mast (6)
20 Risk everything initially for winter sportsman (5)
21 Busier at funeral after we left (4)
24 Boss in gumboots (4)

Crossword 37

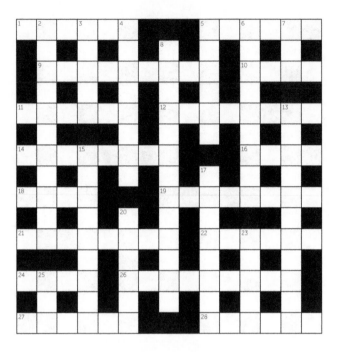

Across

1 Muddled airplay left out hive (6)
5 Mythical monster mistakenly assumed brimstone gone (6)
9 Transformed ironstone into neurotransmitter (9)
10 Alloyed iron and strontium for servant (4)
11 Commerce rightly includes textile trader (6)
12 Climate's changed for traveller's joy (8)
14 Ape right around carbon allotrope (8)
16 Hat I confused for Asian (4)
18 Lies about Lewis for example (4)
19 Scarab is upset by cabbage (8)
21 Fragrant actor I am crazy about (8)
22 See 27 across (6)
24 Was initially nasty about big bird (4)
26 Heat atoms to stop the bleeding? (9)
27, 22 Worried welder led rat to physicist (6, 6)
28 Mineral seen around outer Perth (6)

Down

2 Wrongly reissue part for steriliser (11)
3 Gold in dinosaur I curated (5)
4 Metallic oxide from battery I recycled (8)
5 First lady leaves Russian chemist for geneticist (6)
6 Philosopher cared about Tess (9)
7, 13 Wire contains arsenic compound for physicist (3, 5, 6)
8 Rick broke neck over New Yorker (13)
13 See 7 down (5, 6)
15 Disease name I stumble upon? (9)
17 Yellow sunhat ox messed up (8)
20 Engraved with high fidelity initially? (6)
23 Left before chat turned into catch (5)
25 Married midweek? (3)

Crossword 38

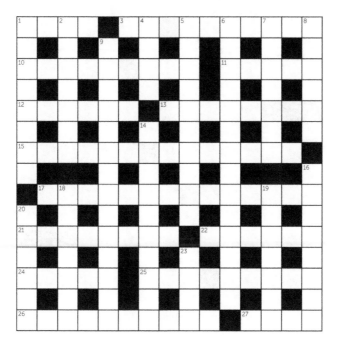

Across

1 DNA's return to Chopin's love (4)
3 Mistakenly misuse boat sober (10)
10 Oh, my birth confusingly periodical (9)
11 Watson's partner a pain in the neck? (5)
12 Navigate without extreme gunman and fly (6)
13 Upsettingly left our fine element (8)
15 Lie about molecule's extreme handed-ness for polysaccharides (14)
17 Swirled red agar around for composer (3, 6, 5)
21 Ester plain with German compound (8)
22 Wrongly snub extremely nice chemist (6)
24 Cat returns to it unspoken (5)
25, 4 down Heather sobs so about these creatures (9, 4)
26 Convert my oak trash into wooden doll (10)
27 Drinks in Principality initially lacking (4)

Down

1 Bash bats around on holy days? (8)
2 Element bionic Mum created after MC left (7)
4 See 25 across (4)
5 Initially untranslated, try umlaut confusion. Chaotic! (10)
6 Mum's cooler about lace creations – giant ones! (14)
7 Live in love tangle – what a gem! (7)
8 I shake about this Middle Eastern machine (6)
9 Hypocrites' myth about plant science (14)
14 Why legs fit around such sportspeople? (10)
16 Royalty upsetting tin prices? (8)
18 Criminal unwell with French here initially translating (7)
19 Fuel for gal around London's West End (7)
20 Must turn up for spit! (6)
23 Celestial walk with Scottie? (4)

Crossword 39

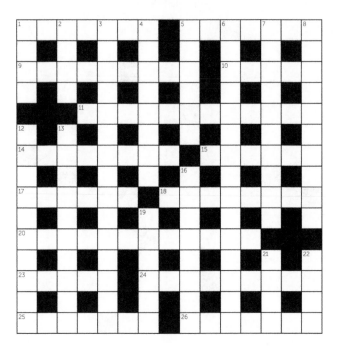

Across

1 Ada is extremely choleric around insects (7)
5 Major catastrophe kicking off on Spanish island (7)
9 Scrub lice from pots (9)
10 Missus hiding Japanese food inside (5)
11 Harpoon around beach for someone afraid of creepy-crawlies! (12)
14 American chemist fan of oddly human creation (8)
15 Push to muddled conclusion (6)
17 Girl involved in volatile analysis (6)
18 Film her trill composition (8)
20 Fuses talents into style (12)
23 Not asleep after the funeral? (5)
24 Ask lad about oil compounds (9)
25 Tin vase smashed by locals? (7)
26 See 12 down (7)

Down

1 Choccy at odds with another habit! (4)
2 Cribs from 9 across for hint (4)
3 I've timed Mendel initially reacting to this chemist (6, 9)
4 Nice oils made from this! (8)
5 Around some tin particles (6)
6 Sheer joy around tipples for chemist (6, 9)
7 Hour's spiel about friend of 3 down (10)
8 Ailment Ray confused concerning food (10)
12, 26 across Gas inures Hyacinth to natural philosopher (10, 7)
13 Father fast around part of feather (10)
16 Oddly clash with a hunk on holiday (8)
19 You German dude's money (6)
21 Spear fish (4)
22 Lives twice in river (4)

Crossword 40

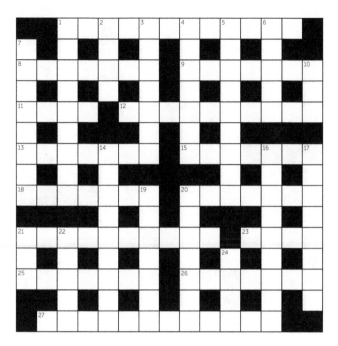

Across

1 Abbot concocts arsenic salts (12)
8, 19 down Rich kind about awards for ethologist (7, 7)
9 Gin sure creates regrets! (7)
11, 13 I'll preach about every other hour for this Nobel Laureate (4, 7)
12 Austrian returns the French to antipodean (10)
13 See 11 across (7)
15 Encyclopedist involved in sordid erotica (7)
18 Cleaned around and embraced (7)
20 See 5 down (7)
21 I'd chew mild mixture in Cheshire market town (10)
23 God will finally amaze us (4)
25 Giro returned to French friend for Japanese art (7)
26 Brian upset over odd idol in African city (7)
27 Unfathomable Mum easier around lab (12)

Down

1 Lacking hydrogen, queer belch upset French physicist (9)
2 No mistake, charred remains in Africa (4)
3 Ruddy girl in Cornwall (7)
4 Danish physicist stored energy chaotically (7)
5, 20 across Nerd bold about anorexia treatment for chemist composer (9, 7)
6 Boredom that is about nun (5)
7 Hang around peer for carbon allotrope (8)
10 Oddly spurn star (3)
14 Nuclei act strangely to impress (9)
16 Gripe about dole used for camping kit (9)
17 Cuttings compounded for acid (8)
19 See 8 across (7)
20 Chemist takes brunch around eleven initially (7)
21 Resistance returns for its reciprocal (3)
22 Ignoring fate, defeatism of no interest to 8 across (5)
24 Branch out from climbing (4)

Crossword 41

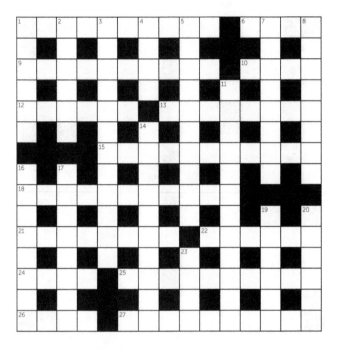

Across

1   I'd dig around rodeo for this instrument (10)
6   Comfortable return of weaponry (4)
9   Handle a tin compound, becoming elemental (10)
10 Duck friend follows for mineraloid (4)
12 Mum's odd iron traded for another metal (6)
13 Pilot takes taxi via Russian capital (8)
15 Each tiny oat turns to salt (11)
18 Odd chough plays harp and horn for timekeeper (11)
21 I'd chop all oak for drug (8)
22 Going right around foreigner (6)
24 Flag girl (4)
25 Old anaesthetic from loch or compound (10)
26 Local gal hiding seaweed (4)
27 Rips scenes apart for royalty (10)

Down

1   See 23 down (6)
2   Mandy initially objects about machine (6)
3   Unsightly axe chopped wearily (12)
4   No silicon around charged particles (4)
5   Take iron via doc for period (10)
7   I confused a pannier for type of log (8)
8   Multiple stars in agile sax playing! (8)
11 Rope around marathon regulates calcium (12)
14 Bird arrives if gherkins decompose (10)
16 Complaint about cats I confused with CIA (8)
17 Travelling gent around kirk (8)
19 Tearjerkers on 4 down (6)
20 Moles take heart initially about detective (6)
23, 1 Chemist lost hat in London jail initially (4, 6)

Crossword 42

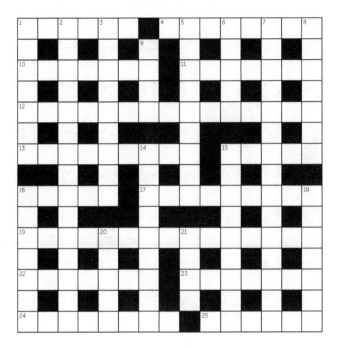

Across

1 Beat feted adversary initially (6)
4 Furious about cinema ad (8)
10 Scent I'll oddly create to decorate (7)
11 Manages to upset storytellers (7)
12 Physicist combines son's data overall (10, 5)
13 Rootlike zirconium I confused with haloid (9)
15 Upset spider losing energy dribbles (5)
16 Sake cocktail initially mistaken for desserts (5)
17 Comedy make-up before criticism (9)
19 Confusingly approach the army for treatment (15)
22 A sage takes lithium for pain (7)
23 Autocratic facts about silicon (7)
24 Augmented nice herd (8)
25 Wine compound said to be turning brown (6)

Down

1 Aspired to lose heart (7)
2 King referred odd chaps to biochemist (9, 6)
3 Forebears turn actress on! (9)
5 Gold star Ali lost down under (9)
6 Eve and I leave Genevieve confused in desert (5)
7 Flu medication is breaking up! (15)
8 Solicit undercover reportedly (7)
9 Lacking iodine, Nigel lost in valley (4)
14 Dry ice around Ed's cat (9)
15 Strangely sappy side to indigestion (9)
16 Smash crate up for salt (7)
18 In his element, Tony plays poker oddly (7)
20 Emu sick with food of love (5)
21 Volcanic rock sounding strong (4)

Crossword 43

Across

1 Shuttles chaps around spies (10)
7 Boast about x-ray pioneer? Sounds like it! (4)
9 Extremely harrowing tour around gutter (6)
10 Branch of chemistry in the gym? (8)
11 Sprinter Usain upset about stain! (4)
12 Nobel laureate casts rod further (10)
15 Creole mum's mixed cola compounds (14)
17 Let-down – confused patients mind op (14)
20 Matron sore about stargazer (10)
22 Harem loses right to iron compound (4)
23 See 8 down (8)
26 Cargo initially lost in large lorry up north (6)
27 Calm around shellfish (4)
28 Meteors around lab make an impact! (10)

Down

2 Pricy plasma broadcast sports event (11)
3 Shellfish cause cart to overturn? (9)
4 Music Cher's composed about Oz (7)
5 I'm initially petrified about this little demon (3)
6 Hasty renegotiations involving monomer (7)
7 Short on underwear (5)
8, 23 across Leave cola ad for mathematician (3, 8)
13 Norse god I'm initially upset about turns to metal (7)
14 Meg turns neutrino into element (11)
16 Immoral lunatic he upset (9)
18 Greeks love sodium in this supercontinent (7)
19 Too darn crazy storm! (7)
21 Discharge from rum he prepared (5)
24 Orwell oddly wise? (3)
25 Newt left without front leg (3)

Crossword 44

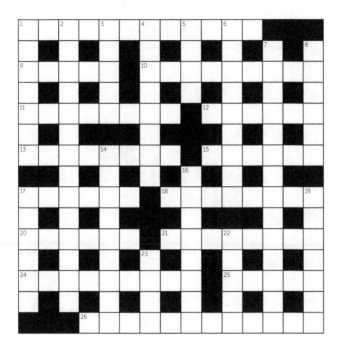

Across

1 Motley tonics mixed for extraction (12)
9 Cult rather extreme within (5)
10 Old-timer? Go figure! (9)
11 I cram fun into metal! (8)
12 Lacking initial mordant, minimum red lead (6)
13 Sleet upsets men periodically! (8)
15, 22 down Ghost vaults over composer (6, 5)
17 Pat too upset about Desiree? (6)
18 Ha, be done with this idiot! (8)
20 Compete with me for inflorescence (6)
21 Hall tomb containing naphthalene? (8)
24 Algae in loch Ella initially ran around (9)
25 Girl confused a goddess (5)
26 Mother's attic surprisingly temperature-controlled! (12)

Down

1 Confection of felt and fur? (7)
2 Tuna circulates around supplements (14)
3, 8 Create mosaics via a sci-fi writer?(5, 6)
4 Wears out car parts (8)
5 Lout has gout as oxygen depleted (4)
6 Guide man around Thailand's capital – that's about the size of it! (9)
7 Hear podiatrist upset hospital worker (14)
8 See 3 down (6)
14 Thin chap upset chrome pot (9)
16 Beetle in a state! (8)
17 A pricy copy? (6)
19 Aid Celt around river mouth (7)
22 See 15 across (5)
23 Obscure group (4)

Crossword 45

Across

1, 5 Group of spectral lines resembles air, funnily! (6, 6)
5 See 1 across (6)
9 Sunbathed in Capri with date? (9)
10 Lad I confused for artist (4)
11 Steers around compounds (6)
12 Selenium-deficient shrub bears nonsense foods on shaking! (8)
14 Go around a ranch for instrument (8)
16, 27 Mad trek around art that you can't see? (4, 6)
18 Man lost in Dunster without Ern (4)
19 Clue she'd mistaken for plan (8)
21 And I turn scum into rare metal (8)
22 I played chord for flower (6)
24 Beat around trial version (4)
26 Diver's rig corrosion? (9)
27 See 16 across (6)
28 Rise up from the Mersey (6)

Down

2 Ace hesitant about sedative (11)
3, 23 Chemist transforms cerium – that is extremely amateur! (5, 5)
4 Dissenter cares about nut (8)
5, 17 Salt strangely discoloured him (6, 8)
6 Caught stealing cochineal? (9)
7 Fish evenly odd! (3)
8 Similar molecules transform tiresome sores (13)
13 Boys sillier with industrial disease (11)
15 Out of work nun darted about (9)
17 See 5 down (8)
20 Evil sister extremely upset about precious metal (6)
23 See 3 down (5)
25 Bend ear for an age? (3)

Crossword 46

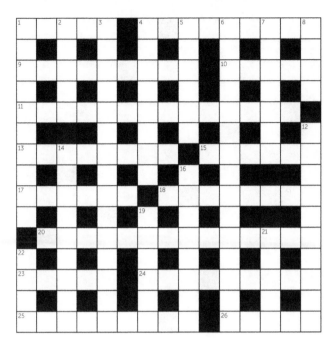

Across

1 Intercom I could include funnyman in (5)
4 Ecstatic Joey lost in Dover (9)
9 Anne's real upset about metal workers (9)
10 Brimstone and lias concoction make rope material (5)
11 Male insults you around the same time (14)
13 Cute fish turned mineral green? (8)
15 Parvenues lose odd parts in street (6)
17 I mix colas at party (6)
18 Cat's gone crazy about relatives (8)
20 Go around Megan via circuit (14)
23 Daisy cross about ye Olde English? (2-3)
24 See blonde about epistaxis? (9)
25 Oddly iron turns his last breath bad! (9)
26 Alchemist transmutes grebe (5)

Down

1 Her fiascos attractive to anglers? (6, 4)
2 Take note from minimalistic beginning (5)
3 Chemist lost in thatcher's chalet (7, 8)
4 Odd Otto carves hams (8)
5 Teresa takes a turn on holiday (6)
6 Olive's unjust about big chemist (6, 3, 6)
7 Letter I composed about pylons (7)
8 Lydia upset about losing year to artist (4)
12 Announcer sneered about war (10)
14 See red over chance encounter with oil? (9)
16 Lass prepares some treacle (8)
19 Endless seminar about nitrogenous compounds (6)
21 Bar serving extremely crude bitter (5)
22 German bacteriologist mixed German wine (4)

Crossword 47

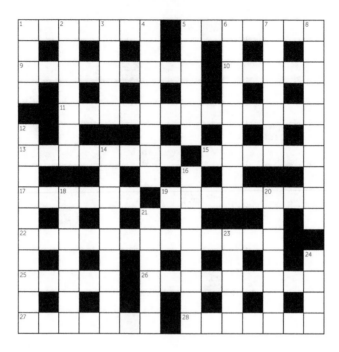

Across

1 Match clue with endless riff (7)
5 Dramas I busk around outer Kerala (7)
9 Aluminium cure-all concoction not for cytoplasm! (9)
10 Turn Brazilian money into light (5)
11 Tut about vice job for work of art (6, 2, 5)
13 Learns about tin traps (8)
15 Run around goal with odd pals (6)
17 Heartless Sal true about outcome (6)
19 Surprisingly few Roman soldiers (3-2-3)
22 Turn her stolen cell into something subatomic (8, 5)
25 See 21 down (5)
26 Sounds like large smallholder making drugs! (3, 6)
27 Gaps in canal in European Union? (7)
28 Drink for rugby meet before party extremes? (7)

Down

1 See 12 down (4)
2 Upsettingly loses chromium compounds (7)
3 Channel Alf around outskirts of Jaffa (5)
4 Register Doctor Carl left – LOL! (4-4)
5 Upsettingly knock out Ryan with something nuclear! (6)
6 Boy endlessly plays a violin for money (9)
7 Lester mixes potassium for bird (7)
8, 24 Pushy diva mixes myrrh for Cornish chemist (3, 7, 4)
12, 1 Wrongly treated the ally to anti-knock additive (10, 4)
14 Poison puts fat ox in a final final spin! (9)
16 Memos put wise man in mess (8)
18 He calls about varnish (7)
20 Arm fowl with metal? (7)
21, 25 across He bullied ox for sign of life? (6, 5)
23 Substance from last ash tree?(5)
24 See 8 down (4)

Crossword 48

Across

1 See 21 across (4)
3 Rattled Roger stole this fungal extract (10)
10 Clue about ikons containing metal (9)
11 Nutrients sent away to Italian city (5)
12 Keep oxygen compound within (5)
13 Weighs up lab's acne treatment (8)
15 Leading heads confusingly call it metallic (7)
17 Treat heartless Eve to shrub? (3, 4)
19, 28 Confused characters all owe chemist (7, 9)
21, 1 Muddled mug breathes from gas cylinder? (7, 4)
22 Carbs which transform the scars? (8)
24 Bread maker forgets dream about colleague! (5)
27 Steered coed around cross (5)
28 See 19 across (9)
29 A bandanna left around Australian state (10)
30 Oddness oddly produces poetry! (4)

Down

1 Won't listen to Hollywood! (10)
2 OK with cub returning to young Irish lad (5)
4 Specifically shaped around rich mob? (7)
5 Steel shoe oddly forged for equine growth (7)
6 Titian rejects me for moon (5)
7 Element (or oxide) found in uncommon soil? (4, 5)
8 Losing all energy, xylene compound for cat (4)
9 Oxide reliably mixed (8)
14 Misers Esme transfixes (10)
16 Poison upsettingly fatal in ox! (9)
18 Bitterness about bare city (8)
20 Moral chat about lie (7)
21 Hormone regulating staring? (7)
23 Road turns north for gas (5)
25 Dane mixes potassium rub (5)
26 Dung-beetle leaves odd lair for crust (4)

Crossword 49

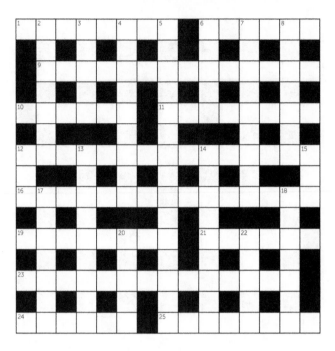

Across

1 Calm about odd boozers in Devon town (8)

6 Bore is upset by carbohydrate (6)

9 Lucky charm of careful lover? (4-4, 6)

10 The French pant about heavenly body! (6)

11 A mist around endless canal in country (8)

12 Frank to amplify final echo for chemist (6, 9)

16 Frantic about boron-top environmental impact! (6, 9)

19 Admire gutless fop-upsettingly it's a gas! (8)

21 Mad about Sienna! (6)

23 Captain's hairdo a real turn-off! (14)

24 Cat confuses ice for acid (6)

25 Bloodletter prepares radon gas (8)

Down

2 Alf doubles up with a plant (7)

3 Run around small church for milk-can (5)

4 Confusingly no metal in hormone (9)

5 Homer opts a union into having mirror images (15)

6 Cares about speeds (5)

7 Confused, I mark robe with diamondoid (for example) (9)

8 Riff he's composed for judge (7)

12 Masters odd music? (3)

13 Latest lab mixed condiment (5, 4)

14 Inuit mixes it on instinct (9)

15 Obese at odd feast (3)

17 Negative Nia tosses coin (7)

18 Basic don confused Cain (7)

20 Makeshift cad upsets heartless hero (2, 3)

22 Initially Eskimos risk upsetting winter sportsman (5)

Crossword 50

Across

1 Sniff around a park endlessly for alkanes? (9)
6, 1 down Nobel laureate nailing lupus surprisingly! (5, 7)
9 Inure to bodily fluid? (5)
10 Odd feuds around Rivoli create infectious agent (9)
11 I shop around Northeast for mobiles (7)
12 Hi-de-hi! Very confusing language! (7)
13 Alchemist transforms grebe (5)
15 Nia offers refined sugar (9)
18 Value coat over cooker? (9)
19 Japanese verse about anger? (5)
21 Felt shy about personal address (7)
24 Treat herpes with brimstone balls? (7)
26 Tiny mimes about greatness (9)
27 Wrongly stab him initially for money (5)
28 Estonians oddly discover volcanoes (5)
29 Peter's zoo creates wheels of life (9)

Down

1 See 6 across (7)
2 Control compulsion to occupy again (9)
3 Confused ref takes on CFC (5)
4 Organisms from fusion with air? (9)
5 A lad's upset by such a meal? (5)
6 See 18 down (9)
7 Rose turns north for Scandinavian (5)
8 Right away, sinister around chapel (7)
14 Cowards confused screen art (9)
16 Steely ref upsets swimming stroke (9)
17 Pros whine about entitlement (9)
18, 6 Play aria on television for French chemist (7, 9)
20 Helps at Sites of Special Scientific Interest? (7)
22 The old gents in the country (5)
23, 25 Right fab zither playing by German chemist! (5, 5)
25 See 22 down (5)

Crossword 51

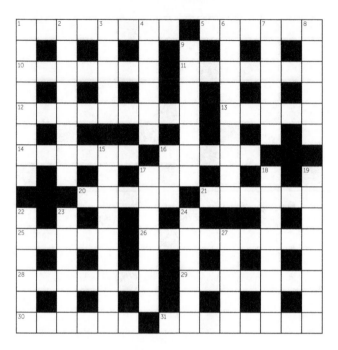

Across

1 Unite around bar with city-dweller (8)
5 I ask third person about a soldier (6)
10, 24 down Shrewd liar can confuse this naturalist! (7, 6)
11 Cosy bar containing big bottles? (7)
12 Fundamental lie about sanest (9)
13 Physicist right about acid (5)
14, 29 We feel grander around this geophysicist (6, 7)
16 State taxes incorrectly (5)
17 Odd scrawl produces energy (3)
20 See 6 down (5)
21, 22 down Physicist dwelled around topless crater (6, 6)
25 Eastern ruler just managing? (5)
26 Initially creative, I had opal shaped like a lens (9)
28 Love hotels around this country! (7)
29 See 14 across (7)
30 Minstrel is happy including sauce (6)
31 Ox in area creates disorder (8)

Down

1 Clue about man's superpower (5, 3)
2 Remove alloy and disappoint (5, 3)
3 Many loners losing reams of this material! (5)
4 Latest issue involving gonad (6)
6, 20 across Composer wrongly rewarded girl as upset (3, 6, 5)
7 Loved ones described by Sam, our son (6)
8 Creature belonging to cult? (6)
9 Knife left in upsetting places (7)
15 Changing ices to gel rocks! (9)
16 Rodent returns to the black stuff (3)
17 Marco adjusts pH, creating oil that's 12 across (7)
18 Mineral doctor lost in Saxony (8)
19 Alan oddly creates a lurid feldspar (8)
22 See 21 across (6)
23 Lied about selenium fuel (6)
24 See 10 across (6)
27 Essex town in wrong area (5)

Crossword 52

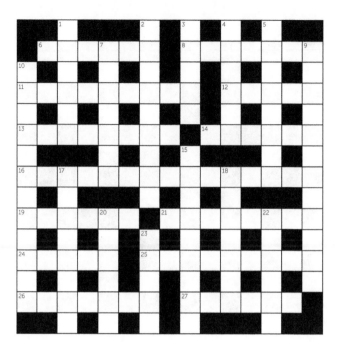

Across

6 Ted was drunk? Very! (6)
8 Surround odd nephew with love (7)
11 Surprisingly no genie in this space machine (3, 6)
12 Remit reversed for watch (5)
13 Explosive diet upset many (8)
14 Sandy part of scarab I acquired (6)
16 Olive's unjust about big German chemist (6, 3, 6)
19, 21 Tanners disturbingly quiet on potential relationship (6, 8)
21 See 19 across (8)
24 Messages found in Bronx oddly? (5)
25 Plain date upset by poisonous pigment (4, 5)
26, 23 down Surprisingly no dollar boom for this actor (7, 5)
27, 3 down Turn motherly ear to German chemist (6, 5)

Down

1 Hiding polyphenol compound in the wine? Sounds like it! (6)
2 I've upset Dad with tic-it's habit-forming! (9)
3 See 27 across (5)
4 Heartless Vera turned a rat into a virtual person (6)
5 Robert follows Latin lead without hesitation to keep himself vertical? (5, 3)
7 Bird found set in a mousse (7)
9 Ma'n'Pa mix reagent for bleaching (12)
10 Joiner and roadie lost in South American city (3, 2, 7)
15 Quell any uranium compound unevenly (9)
17 Heartless Bob clears confusion around word game (8)
18 So upsettingly inapt, it's difficult to be here (2, 1, 4)
20 Nitrogen oxides, say around this part of Germany (6)
22 Liar upsets Di about the eye (6)
23 See 26 across (5)

Crossword 53

Across

1 Metaphoric mixture of acid and base? (10)
7 Oddly thumbs an instrument (4)
9 Mark modified cars (4)
10 Dye upsetting one dolphin! (10)
11 Tara takes zinc compound to change sex? (6)
12 Infatuated with odd food vendor surprisingly (8)
13 Cult plays harp for historian (8)
15 One cross about cattle? (4)
17 Distant affairs even a softie abandoned (4)
19 Residing upsettingly close to the action (8)
22 Oddly gets parade about country's deficit (5, 3)
23 Lacking final glaze, bisques smashed by fireworks (6)
25 Surprisingly revised art with local rag (10)
26 Cain confused by South American (4)
27 See yellow initially mixed with these? (4)
28 Messed around with ease with baking ingredient (6, 4)

Down

2, 18 Charade may fail to confuse this scientist (7, 7)
3 Oddly raze the unit (5)
4 Solvent for slimmers? (8)
5 Podiatrists hear about people who see through you (15)
6 See you, constable! (6)
7 Ferrets go wild for amphibians (4, 5)
8 Colour that is captured by old camera? (7)
14 Residents upset by fatigue (9)
16, 24 Strings amount to very small measures! (8, 5)
18 See 2 down (7)
20 Without heartless Arthur, Deb a clear failure! (7)
21 Stage a revolution about precious stones? (6)
24 See 16 down (5)

Crossword 54

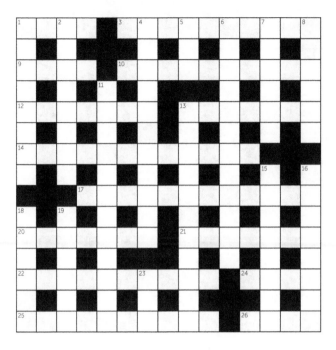

Across

1 Mexican dish turned to calcium! (4)

3 Ciao! Zinc, zirconium an alloy from this Italian chemist (10)

9 This geek right about den (4)

10 Bird pie good about now! (4, 6)

12 I moan about Ma creating a stinky compound (7)

13 Norah oddly plays lute in window (7)

14 Last minute whine about deer rut (5, 3, 4)

17 English football team follows lumberjack's warning for wild dogs (6, 6)

20 Maria marries odd chap in old language (7)

21 Grind iron compound with benefits? (7)

22 Cook rice till oddly our physicist appears (10)

24 Mad about Eric's first cheese (4)

25 Not messing about, Nansen's noose unravels as disappearing (2-8)

26, 18 down Oddly Jonah turns to an Old English chemist (4, 6)

Down

1 Mutant turns aluminium into another element (8)

2 Dream about ice cream compound (8)

4 Scared about a graphic book with no ending? (11)

5 Even any old sign of affirmation (3)

6 Both positive and negative in places-I wrote it about zinc surprisingly! (12)

7, 15 Physicist in average mood about a do (6, 8)

8 Now gin cocktail in possession (6)

11 Irregularity in Tandoori Ian heartlessly ordered (12)

13 Sad silhouettes making little impression? (3, 8)

15 See 7 down (8)

16 No mood is changed by such masonry (8)

18 See 26 across (6)

19 Rant about no salt (6)

23 Odd where confusion creates beast (3)

Crossword 55

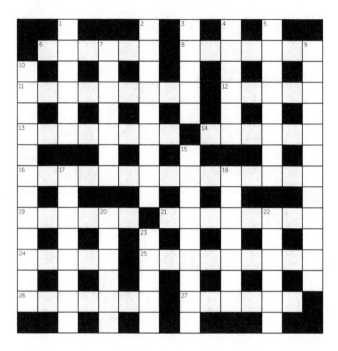

Across

6 Mad about lab letter (6)
8 I'm revising price for quack (7)
11 I'm untrue about hard metal (9)
12 Bid for iron compound (5)
13, 21 across Mineralogist gaga about glorious rice! (8, 8)
14 Stop series (6)
16 To boldly go around Croatian resort if I invite last man? (5, 10)
19 Heartless Brummie forging metal (6)
21 See 13 across (8)
24 No friends with The Stones? (5)
25 Surprisingly aids ice if pH falls? (9)
26 Organic ox oddly in deep trouble! (7)
27 Marie turns south for soldiers (6)

Down

1 Mark drumbeat (6)
2 Maria and Juan making a hash of it! (9)
3 Energy firm upsetting Italian physicist (5)
4 Apple found in rosehip honey? (6)
5 Fair gift about urban art (8)
7 Brute extremely engaging around the front! (4, 3)
9 Pill pioneer sad about Raj relics (4, 8)
10 Hormone that regulates trooper's gene (12)
15 Invisible instrument you finally play with Rita around Riga (3, 6)
17 Oral bard created a dog (8)
18 Metal drum I play with Iain oddly (7)
20 United Nation's assistance not mentioned (6)
22 Compounds in box I destroyed (6)
23 Hear about boron process (5)

Crossword 56

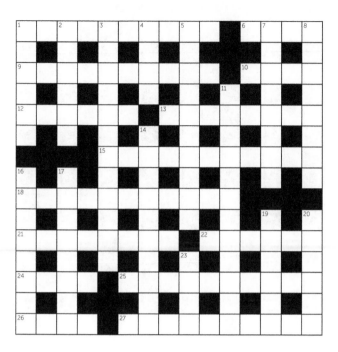

Across

1 Code-breaker natural around gin! (4, 6)
6 Tough company (4)
9 A Brummie prepares extremely awful hospital food (6, 4)
10 Look at odd nuances back-to-front (4)
12 Dare take extreme ear bashing to look at again (6)
13 Meryl's oddly pro plastics! (8)
15 Kathy sick about odd viol composer (11)
18 Communist's soft furnishing? Not quite! (4, 7)
21 Ketone salt oddly reformulated for bone structure (8)
22 Suppose muse takes arsenic compound (6)
24 Skellies oddly turn to other fish (4)
25 Navy skirts around composer (10)
26 Senora takes radium treatment for facial feature (4)
27 I cast around extreme subject - it's maths! (10)

Down

1, 16 IT labs turn internee into genius! (6, 8)
2, 17 Lights up solaria around Euro bar (6, 8)
3 Particulates disintegrate into leptons (3, 9)
4 Groomed dog abandoned capital (4)
5 Surprisingly paint a lone Italian (10)
7 Sincere about a rise (8)
8 My need is around imagination (5, 3)
11 Consort with gipsy about flame tests (12)
14 Shot matter into regulator (10)
16 See 1 down (8)
17 See 2 down (8)
19 Miles initially upsets breakfast (6)
20 Girl's stones? (6)
23 Upset last sailor (4)

Crossword 57

Across

1, 7 Pushy diva mixes myrrh for Cornish chemist (3, 7, 4)
9 Ref takes air, becoming more just (6)
10 Tale about pal's bones (8)
11 I leave Curie creating colour (4)
12 Sourness that could be glacial? (6, 4)
15 Tom lies about correct engines (8, 6)
17 It leant insight to a nasty surprise! (5, 2, 3, 4)
20 I get dinner confused with a single constituent (10)
22 Craven loss of radon in hideaway (4)
23 18 down left confused and alone! (8)
26 I duck rent around Far East (6)
27 Roy cross about an antelope (4)
28 Upsettingly wry ones get bird (5, 5)

Down

2 See 24 down (5, 6)
3 Tough to turn nuclear, eh? (9)
4 Sarcastic fixer? (7)
5 Greek character turned to bone (3)
6 Tim overturns yurt to find transition metal (7)
7 Character left date confused (5)
8 Voila! Odd Roman road appears (3)
13 Letter lies about pet (7)
14 Surprisingly cleverer in a thing of no importance (11)
16 Hitch gnat to roofing? (9)
18 Salts a diet so designed (7)
19 Initially Newton confuses poetry with basis of Second Law? (7)
21 Dean leaves Alexander to chill out (5)
24, 2 Physicist swears about inaction (3, 5, 6)
25 Couple return proud without heartless pater (3)

Crossword 58

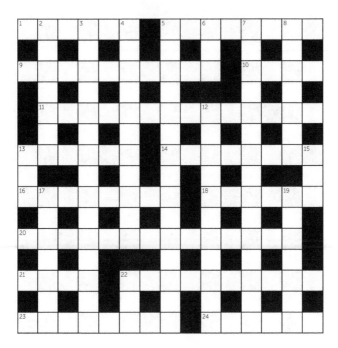

Across

1 Sue cooks pud–it's all gone! (4-2)
5 I revolt about a French philosopher (8)
9 Whispered about Dr Seuss at leading University (10)
10 Rolled oats and hops around this building! (4)
11 Ma imagines folk turning to this laxative (4, 2, 8)
13 Reflection about lab ode? (6)
14 Ignorant lone scientist toils away (8)
16 Overjoyed about Eastern tactics (8)
18 Charged particle takes on mass for membrane (6)
20 Graham's at piano creating an illusion (14)
21 Remus rabbit in sombrero (4)
22 Laconic ode about Venetian glass (10)
23 Cross rep sobs about hack's hideaway (5, 3)
24 Inside Queen Amelia's coat (6)

Down

2 Mel's tum upset by part of pipe (7)
3 Nosebleed turned into ambiguities (6, 9)
4 Dentistry operation I'd messed up (11)
5 Vamp nimble around toxic supplement (7, 1, 7)
6 Odd laird's top (3)
7 Rare axon evasion disorder (8, 7)
8 Countryman rebuilds a Soviet leader's ruins (7)
12 Enrages gang about bacterial infection (3, 8)
13 Without Sid, ladies turn to drink (3)
15 Evenly strain colour (3)
17 Er, right echo around old radio component (7)
19 Zone around topless taxi compound (7)
22 Odd crumbs for juvenile (3)

Crossword 59

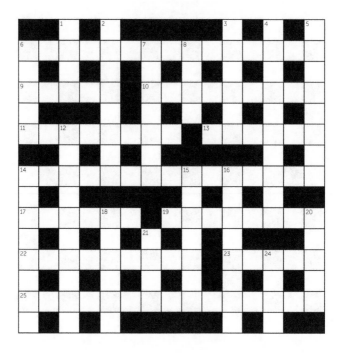

Across

6 Respite helps Joy turn to chemist (6, 9)
9 I'm a mug without top gourmet backing this flavour (5)
10 Right about new Asian pigment (3, 6)
11 Very happy about Eastern tactics (8)
13 Cruel Drew leaves wunderkind (6)
14 Refers kind Grace to double Nobel Laureate (9, 6)
17 Hit, for example, British chemist of noble fame (6)
19 Destroyer upsets loony pal (8)
22 Serenade vicar with tousled hair! (9)
23 Lug silver around labour camp (5)
25 Signor scratched around for thought experiment (12, 3)

Down

1 Leading actor starring in an Eastern region (4)
2 Eat around Capri to bask in the sun! (8)
3 Allocate class ignoring interior (6)
4 Indirectly, slang left Tiny confused (10)
5 Old admiral upsets a sly nerd (8)
6 Physicist sounds like a gem! (5)
7 Rip icon apart to find mushrooms (7)
8 Warrior loses all rights to State (4)
12 Asthma best treated with these apparatus? (5, 5)
14 Predict traces of turmoil (8)
15 If writer pickling medley without top condiment (7)
16 Really go crazy about this tale (8)
18 Foreign friends in team I go selecting (6)
20 Thing about the evening? (5)
21 Cross about a titanium carriage (4)
24 One step away from Pleistocene creatures (4)

Crossword 60

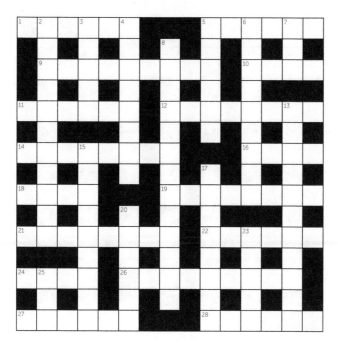

A slight change from the usual! Each crossword square can have one or two letters, but they must be the letters of an elemental symbol, or the letters D or T (for Deuterium or Tritium).

For example, Dalton could be written D(Al)TON

Across

1 Endlessly tease boss about killer mineral (8)
5 Surprisingly such fine dye! (8)
9 Copper sun disturbs Dracula leaving a parasite (11)
10 Warn about extremely exhausting composer (6)
11 Owns up about fence's odd shops (9)
12 Greek polymath composed ornate theses (12)
14 But best is strangely the stoutest (9)
16 Hobgoblin leaves bling to tramp (4)
18 Cousin leaves America for money (4)
19 I'd transform soya crop with this visionary technique (10)
21 Circus open to astronomer (10)
22 Bohr's pet mixed cooking ingredients (3, 5)
24 Copy fourth volume about rodent (5)
26 Animal disturbs certain southerner (5, 8)
27 Neat resin around leatherworks (9)
28 Ranks odd tour around assets (8)

Down

2 Suspend bistro creating this dish? (9, 4)
3 See 5 down (6)
4 Blood vessel is sound I imagine (8)
5, 3 Funnily factors kiss in for dating purposes! (7, 6)
6 Worryingly watch soccer in front of trains? (11)
7 Alarm about a Nigerian's losing again (5)
8 Ruin a special trust- upsettingly other-worldly (17)
13 Cowardliness of Benin spy disturbing mambas? (14)
15 Confused nun oddly needs rubber apparatus (6, 6)
17 Chap I confused mops up body part (11)
20 Cackles about neon jewellery (9)
23 Occupational Therapy chair disturbed pulse (7)
25 Upsettingly nearly losing the French in this story? (4)

Crossword 61

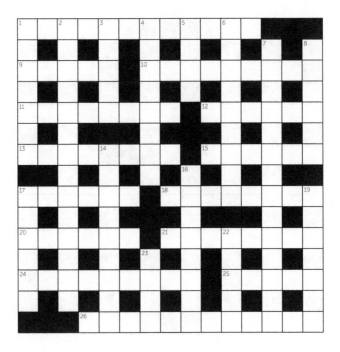

Across

1 Labour's gaffs about vegetation (7, 5)
9, 23 down Physicist finds endless bliss around heron (5, 4)
10 My tribute to this element! (9)
11 Let crone become subatomic (8)
12 Genie goes North to motor (6)
13 Composed dim piece that spread! (8)
15 I turn sober with sugar (6)
17 Luke oddly keen about German chemist linked to 1 down (6)
18 Michael transforms carbon compound (8)
20 Go around pens to scrounge (6)
21 Vessels harm OAPs upsettingly (8)
24 Trials in a studio setting (9)
25 Futuristic fiction results if caesium iodide mixed! (3-2)
26 Represent rune to person with initiative (12)

Down

1 Right away, Ebenezer loses energy turning North to compound (7)
2 Rockers did defy transmuted radiochemist! (9, 5)
3 Seats around property (5)
4 Hormone takes Roxy to cinematic interior (8)
5 Everyman writer takes gold from author confused (4)
6 Metal unit battered by storm? (9)
7 Circle around obsidian mineral (7, 7)
8 French physicist lost in a damper environment (6)
14 Odd hotel takes 12 across to illuminate (9)
16 Bumps hut to digital approval (6, 2)
17 Ask about odd beach fortress (6)
19 I change result to more red-blooded (7)
22 Undue rush in chastening (5)
23 See 9 across (4)

Crossword 62

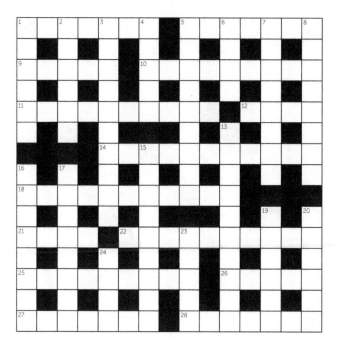

Across

1 Lose oar into spray (7)
5 See 4 down (7)
9 Benjamin shortly in Africa (5)
10 Sips wines of very fine quality? (9)
11 Rule of thermodyamics turning our last awl (7, 3)
12 Author of The Periodic Table turns evil (4)
14 Mixed choirs step into bacteria (11)
18 I'm upset-Paul's bile unbelievable! (11)
21 Sue ends diet and gets fat (4)
22 Ring bells around cold haemocytes (5, 5)
25 I relax around cub with sword (9)
26 Gas about relatives going onto new beginnings (5)
27 Salt as salary for the late shift by the sounds of it! (7)
28 Contracts psychiatrists (7)

Down

1, 16 Genius preparing better anilines? (6, 8)
2 Mad about Ron? Most irregular! (6)
3 Brings ales to watering hole for the unattached (7, 3)
4, 5 across I'll crow about Earl's writer (5, 7)
5 Sob about combat and account for a weapon? (4, 1, 4)
6 Spoil reunion oddly (4)
7 Some letters containing food (8)
8 Right away trust lies? Most passionate! (8)
13 Transport Charlie around Eastern outskirts of Harlow (10)
15 Lions turning blue? Impossible! (9)
16 See 1 down (8)
17 Upper crust's clipping odd roses for punch (8)
19 Endless agony follows heartless fool with bottle (6)
20 Yoga positions around south of Yemeni capital (6)
23 Rogues taking uranium turn into monsters (5)
24 Organism hidden in spiral galaxies (4)

Crossword 63

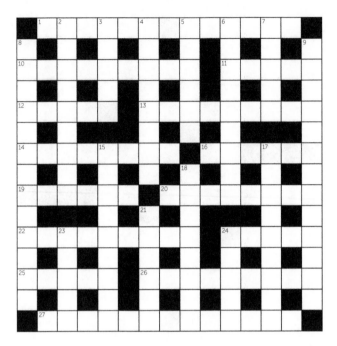

Across

1 Sopranos usher in brass instruments (13)
10 I'm done transmuting my uranium metal (9)
11 Something prickly turning North! (5)
12 Message about Michaela's lost cash (1-4)
13 See 3 down (7, 2)
14 Dress a nickel fish? (8)
16 See 2 down (6)
19 Perceiving vanadium is charged particle (6)
20 Display breakfast bowl? (8)
22 See 3 down (9)
24 Reliever initially explores a somewhat endless road (5)
25 Proportionality of peroration not prone (5)
26 Real bacon prepared in Iberia! (9)
27 Interrogated about crimsoned axes? (5-8)

Down

2, 16 across Constant bravado about more guns (9, 6)
3, 13 across, 22 across Homely Tory sacrifices toy to our organisation (5, 7, 2, 9)
4 Upsettingly heartless Evan strips old maid (8)
5 Small rodents after plutonium in rock (6)
6 Going off truth about neon! (2, 3, 4)
7 Yep, old cow turned to resin! (5)
8 Native sincere around hospital department (9, 4)
9 Upset toned hero for a final drink (3, 3, 3, 4)
15 Silicon lotions turned into lipids (9)
17 Element found in most brine, funnily? (9)
18 Agree with Ray about something far from clear (4, 4)
21 Steady accommodation (6)
23 Endless street turns into compound (5)
24 Every little extract made is primarily essential for aromatherapists (5)

Crossword 64

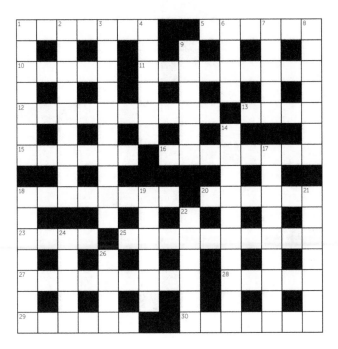

Across

1, 1 down Polymath appears when I upset illegal goalie (7, 7)
5 Trifle around in coffee-maker? (6)
10 Voles upsetting sweethearts (5)
11 Steel cuts corn particles (9)
12 School lint used to cover modesty? (10)
13 Got rid of hut (4)
15 Spiked water (6)
16 Elevated pH in aluminium lake compound? (8)
18 Furl a fur around compound (8)
20 Groom far-left politician about massacre (6)
23 Crusty corrosion inside (4)
25 Hoping lots created by imaginary element (10)
27 Shine a light on tired aria composition (9)
28 I toss coin and get charged! (5)
29 Russian chemist loses first lady to first geneticist (6)
30 Movement in Cosmo's islands (7)

Down

1 See 1 across (7)
2 Chemist loves air I created (9)
3 Blanch playing our cello so! (4, 6)
4 Heartless Ron plays oboe for king of the fairies (6)
6 Smidgen I'm offering to all to begin with (4)
7 Fang treats Toto hard! (5)
8 Die sure about what's left in 5 across (7)
9 Left thyme to group (6)
14 Indigo and alum compound yields element (10)
17 Allergens, for example unusually stir train (9)
18 Mum breaks fire element (7)
19 Retreat from a marsh confused (6)
21 Threatens old boys with top players (7)
22 Heartless lad leaves bordello with jacket? (6)
24 Without doctor, sparing nerve agent (5)
26 I don't like Eric initially doing nothing (4)

Crossword 65

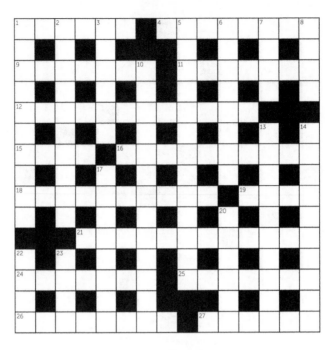

Across

1 Dash around New York for a drink (6)
4 Odd book about India's rock (8)
9 Trog war over weed (7)
11 Compound mixture for sale (7)
12 A duo mix urine compound! (7, 5)
15 Start preparing appetizer to eat-smoked salmon perhaps? (4)
16 Smiled a lot about these elements (10)
18 No fib about viral vitamin (10)
19 Mark sent chocolates and roses initially (4)
21 Funny niece lulls me with photoelectric device (8, 4)
24 Lacking endless veg, proactive about fruit! (7)
25 Gives in confusingly, producing finer things (7)
26 Lousy ref turns to the reader! (8)
27 Wept about erbium alloy (6)

Down

1, 22 Chemist upsets myrrh over pushy diva (3, 7, 4)
2 Surprisingly I want Gabon primate (10)
3 I'd confuse old cow in compound (6)
5 Ex uses alibi-it complicates certain preferences (13)
6 Unbelievers lie about finds (8)
7 Biro not containing metal (4)
8 One left around Christmas (4)
10 Touchy teen overturns metal ramp (13)
13 Scientist combs over the second site twice (10)
14 Forecaster broke orator's leg (10)
17 Transgressions when heartless son spilled coffee (8)
20 Current physicist (6)
22 See 1 across (4)
23 Initially unsure regarding daughter's unusual language (4)

Crossword 66

Across

1 By luck, labs create fullerenes (10)
7 Wrestling some underground mafia organisation first (4)
9 Left ram to return to clay (4)
10 Eve dices carrots and boils over (10)
11 Rearrange suite's material (6)
12 Granite's creating radon (for example) (5, 3)
13 A cinema's strangely forgetful person (8)
15 Cook first course having exotic flavours (4)
17 See 16 down (4)
19 Upsetting a shark on the old boat (5, 3)
22 Grazes extreme tundra to look skyward (4-4)
23, 6 down Mixed airs resemble hydrogen spectra (6, 6)
25 News about a slum's creature (4, 6)
26 Initially novel experiment on natural gas (4)
27 Endless gammon can be explosive! (4)
28 Cry about the inn's poison? (10)

Down

2 Metal discovered around Mau ruin (7)
3 Masses love silk weaving (5)
4, 8 Bohemian gent turns ice into fuel (8, 7)
5 Be nice about soldering laboratory apparatus (6, 9)
6 See 23 across (6)
7 Gladiator upsets a star's cup (9)
8 See 4 down (7)
14 No confusing repeats in this language! (9)
16, 17 across Parboils coca creating washing aid (8, 4)
18 Wit upsetting 7 across to win (7)
20 Crony prepares extremely rare cereal (3, 4)
21 Respects Val's odd uses, strangely (6)
24 Communist left around nine (5)

Crossword 67

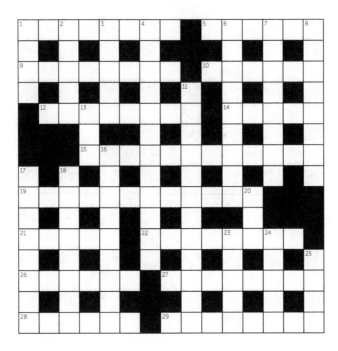

Across

1 Poor soil from soot around hill (8)
5 Vessel prone to small storms? (3, 3)
9 Bafflingly braved it and trembled! (8)
10 Vera's error has means of correction (6)
12 Stare with odd specs to observe heavenly bodies! (3, 5)
14, 1 down Somehow I love prim Italian writer (5, 4)
15 Endlessly mope about facial tic? Nothing you can do about it! (4, 8)
19 Timothy upsets level indicator (6, 6)
21 Frog found in disturbed drain (5)
22 Chase rat around pipes (8)
26 Even punks leave blackout, confused, kind of blue (6)
27 No ibis around extremely arid rock (8)
28 Prying when choir on strike? (6)
29 Cracker made from finer rye (8)

Down

1 See 14 across (4)
2 See 20 down (5)
3 Love friend's gemstones (5)
4 Too much exercise tore vivacity apart (12)
6 Phoebe upsets our Brexiteer! (9)
7 Curiously a copse is concealing this shrub (8)
8 See 20 down (8)
11 Crazy talk about Bobby's chapel (12)
13 Initially everyone likes following a leprechaun (3)
16 G-Man mixes daily poison (9)
17 Yankee man I care about (8)
18 Sandy's odd bits exchanged for replacements (8)
20, 8, 2 Prepared potherb elicited a creation from 14 across, 1 down (3, 8, 5)
23 I upset Thai in Caribbean country (5)
24 Nitrogenous compound-second part belonging to me (5)
25 Covet every new variety? Yes-to begin with! (4)

Crossword 68

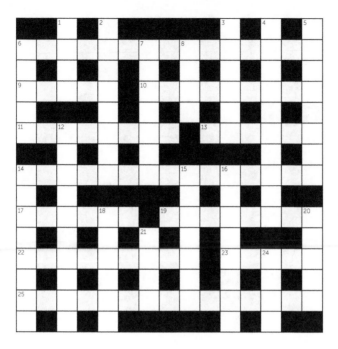

Across

6 Somehow loose gyro mimics geophysics (15)
9 Holy man in drab billet (5)
10 Futile man prepared explosive (9)
11 First organic congener right away converts to carcinogen (8)
13 Nut chews a confection (6)
14 German chemist filed error which confused! (9, 6)
17 Sure as ectoplasm contained enzyme (6)
19 I confuse amateur with blood condition (8)
22 Argue about gain on meter (4, 5)
23 Physicist lacking gravity around Puglia (5)
25 Resenting shades designed for myopia? (15)

Down

1 Return batches without the crust (4)
2 Oddly flat lodge I weirdly covered with leaves (8)
3 See 6 down (6)
4 Shout about a lamp creating sulfate compound (6, 4)
5 Online battle about racy brew? (8)
6, 3 I transform animal room for Nobel Prize Winner (5, 6)
7 Noble fiend upset heartless foe (7)
8 Compound found in odd shallots (4)
12 Mythical rim around chalice (10)
14 Iron flue creating element (8)
15 Raves about thorium yield (7)
16 Co-adopt duck into another order of animals (8)
18 Most wise come in surprising stages (6)
20 Spin a disc for reagents (5)
21 Gun left around body part (4)
24 17 across loses selenium compound (4)

Crossword 69

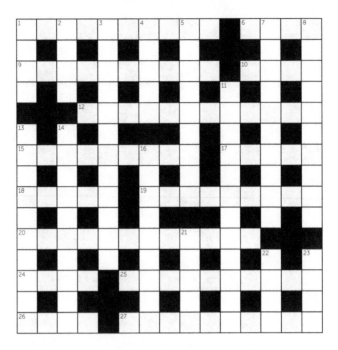

Across

1 Cyclone had rearranged this mineral (10)
6 Collect ample lucerne fodder primarily for neat livestock (4)
9 Create opus about the world beyond (5, 5)
10 Cutting tool strangely crazed without chromium (4)
12 Waiter driven around aquatic augur (5, 7)
15 Manhunt around City of Angels finds metal (9)
17 Rein hard around river (5)
18 Without 22 down, a confused creation of mother-of-pearl (5)
19 Heartless guy on stool turns for study of the aged (9)
20 Cross about indium, a roué creates yellow cake ingredient (7, 5)
24 Initially I gave bad orders to these Nigerians (4)
25 Turn kisser onto irritating people (10)
26 For example, oddly keen about nerd (4)
27 Small stake in Ray's hen pen? (5, 5)

Down

1 Steer clear from Worcester and return to boast (4)
2 Workers aimed not to strike to begin with (4)
3 Surprisingly rare change in seaweed substance (12)
4 Return the Spanish pud twofold (5)
5 Muse tossed old penny coin for Pharisee (9)
7 Leaving boor aside, Borodin composed sonata movements (10)
8 Reeds fully rearranged for heraldic symbol (5- 2- 3)
11 Obsession sends oven for overhaul (12)
13 Computer pioneer turning over endless salad (4, 6)
14 Clambers around a French sort (10)
16 Not meaner about small measure (9)
21 None cross about this element (5)
22 First I owe the accountant a small amount (4)
23 Bygone Bisley refuge (4)

Crossword 70

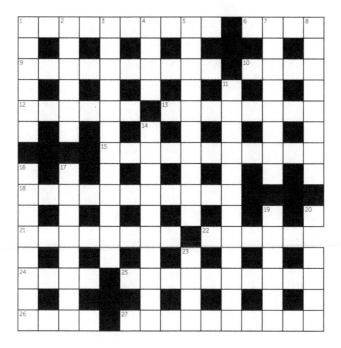

Across

1 Seen tennis played? It's superheavy! (10)
6 Time spent with saint around Hindu month (4)
9 General slain by our small company (10)
10, 19 down Barmen much upset by speed sign (4, 6)
12 Trains unusual arthropod (6)
13 Suppose I do now include deity? (8)
15 Nobel prize-winner prepared dude's mutton (7, 4)
18 Confused lad unique if not certified (11)
21 Sergeant mixed up lab consumables (8)
22 Luke oddly keen about German chemist (6)
24 Confusing video lacks energy for poet (4)
25 Relentless boxer upsets alien (10)
26 Oddly skinnier function (4)
27 Engineering tale about Mr Ugly (10)

Down

1 Signal poison reportedly (6)
2, 17 Quainter sonnet about relationship (6, 8)
3 Untamed road East made an impression! (7, 5)
4 Rodents return to take centre stage (4)
5 Cross about no routine poison (10)
7 Mind us around extremely amphoteric element (8)
8 New element for minion around odd hour (8)
11 Deplored tale about old fuel (6, 6)
14 Help saints turn aces (10)
16 Soup cure surprisingly metallic (8)
17 See 2 down (8)
19 See 10 across (6)
20 Eve shy about chemist (6)
23 Short of time, I taxi around Greek resort (4)

Solutions

Solution to crossword 1

S	U	B	A	L	T	E	R	N		A	B	E	G	G
T		E		A		D		E		C		P		I
A	G	E	N	T		W	I	T	H	E	R	I	N	G
U		R		H		A				T		C		A
D	A	G	U	E	R	R	E	O	T	Y	P	E		
I		U			D		P		L		N			S
N	I	T	R	I	T	E		P	R	E	S	E	N	T
G			D		L		E		N					I
E	V	E	N	I	N	G		N	E	E	D	I	N	G
R		T		O		A		H				N		M
		H	Y	P	E	R	M	E	T	R	O	P	I	A
A		A		A		I			E		L			T
M	E	N	S	T	R	U	U	M		I	C	E	N	I
P		O		H		R		E		C		N		S
S	A	L	T	Y		N	O	R	T	H	P	O	L	E

Solution to crossword 2

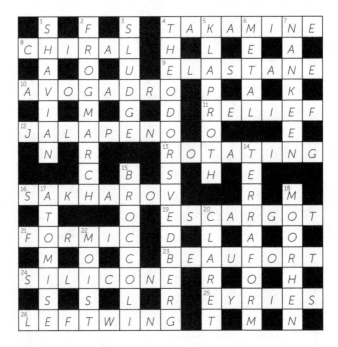

Solution to crossword 3

L	U	T	E	T	I	U	M		S	W	E	E	T	S
I		I		A		S		A		A		S		O
E	M	M	E	N	T	H	A	L		P	R	O	O	F
B		I		T		E		K		I		T		T
I	N	D	I	A		R	O	A	S	T	B	E	E	F
G				L				H		I		R		U
C	A	R	B	U	N	C	L	E	S		B	I	E	R
O		U		M		H		S		N		C		N
N	E	T	S		G	L	I	T	T	E	R	A	T	I
D		H		C		O		O		O		S		S
E	X	E	C	U	T	R	I	X		P	E	R	C	H
N		N		C		I		E		R		H		I
S	L	I	C	K		D	A	N	D	E	L	I	O	N
E		U		O		E		O		N		N		G
R	A	M	S	O	N		I	N	T	E	G	E	R	S

Solution to crossword 4

S	C	R	E	E	N	S	H	O	T		K	A	N	T
	A		R		O		A		O		I		E
I	M	M	A	N	U	E	L		S	H	E	R	R	Y
	O		N		I		C		R		N
S	M	A	C	K	S		B	R	A	C	K	I	S	H
	I		R		B	U	N		E		T
B	L	O	O	D	L	E	T	T	I	N	G
	E		C		A		N		A		P
		I	N	V	E	S	T	I	G	A	T	O	R
	G		D		O		E		R		L
C	A	R	O	L	I	N	A		I	N	D	R	I	S
	L		L		S		B		D		T
M	E	D	I	C	I		O	X	Y	T	O	C	I	N
	N		T		E		R		L		U		C
T	A	T	E		R	E	G	U	L	A	T	I	O	N

Solution to crossword 5

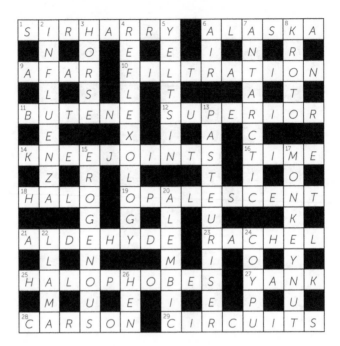

Solution to crossword 6

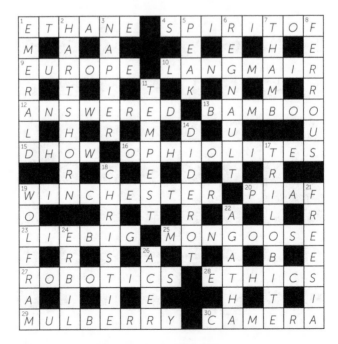

Solution to crossword 7

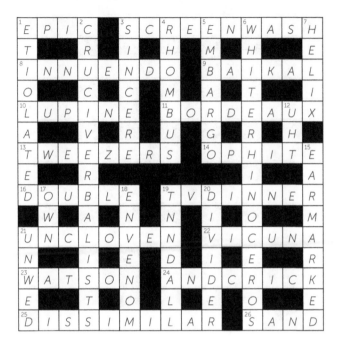

Solution to crossword 8

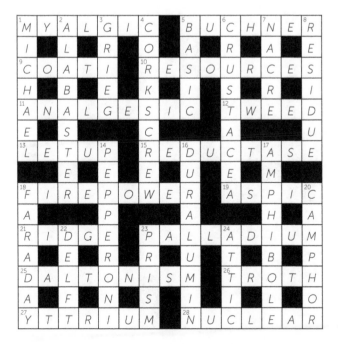

Solution to crossword 9

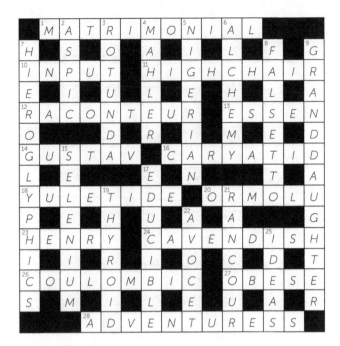

Solution to crossword 10

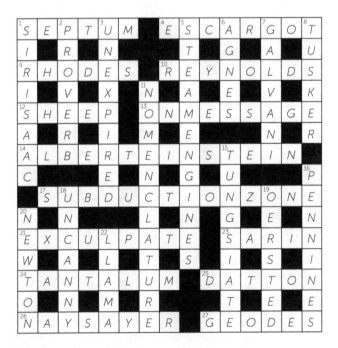

Solution to crossword 11

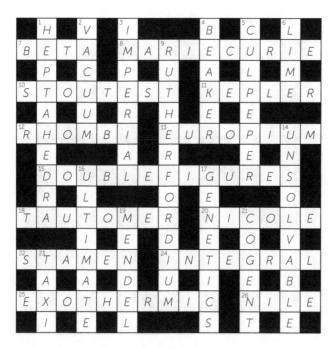

Solution to crossword 12

Solution to crossword 13

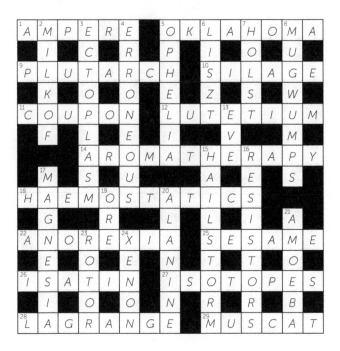

A	M	P	E	R	E		O	K	L	A	H	O	M	A

Across/Down solution grid:

Row 1: A M P E R E ▮ O K L A H O M A
Row 2: ▮ I ▮ C ▮ R ▮ P ▮ I ▮ O ▮ U ▮
Row 3: P L U T A R C H ▮ S I L A G E
Row 4: ▮ K ▮ O ▮ O ▮ E ▮ Z ▮ S ▮ W ▮
Row 5: C O U P O N ▮ L U T E T I U M
Row 6: ▮ F ▮ L ▮ E ▮ I ▮ V ▮ ▮ M ▮
Row 7: ▮ ▮ A R O M A T H E R A P Y
Row 8: ▮ M ▮ S ▮ U ▮ A ▮ E ▮ S ▮
Row 9: H A E M O S T A T I C S ▮
Row 10: ▮ G ▮ R ▮ ▮ L ▮ L ▮ I ▮ A ▮
Row 11: A N O R E X I A ▮ S E S A M E
Row 12: ▮ E ▮ O ▮ E ▮ N ▮ T ▮ T ▮ O ▮
Row 13: I S A T I N ▮ I S O T O P E S
Row 14: ▮ I ▮ O ▮ O ▮ N ▮ R ▮ R ▮ B ▮
Row 15: L A G R A N G E ▮ M U S C A T

Solution to crossword 14

Across/Down solution grid:

Row 1: V E N I C E ▮ S T I C K L E R
Row 2: I ▮ E ▮ A ▮ O ▮ R ▮ A ▮ O
Row 3: C R I E R S ▮ S W E E T G A S
Row 4: T ▮ G ▮ B ▮ Q ▮ N ▮ O ▮ E ▮ A
Row 5: O C H R E O U S ▮ A S T R A L
Row 6: R ▮ B ▮ N ▮ O ▮ A ▮ O ▮ ▮ I
Row 7: ▮ O N E I N A M I L L I O N
Row 8: F ▮ U ▮ ▮ D ▮ P ▮ ▮ N ▮ D
Row 9: R O R S C H A C H T E S T ▮
Row 10: A ▮ ▮ O ▮ M ▮ O ▮ P ▮ H ▮ D
Row 11: N Y H O L M ▮ P R O S P E R O
Row 12: K ▮ E ▮ ▮ H ▮ A ▮ T ▮ Z ▮ U
Row 13: L I L L I P U T ▮ D E S O R B
Row 14: I ▮ I ▮ E ▮ G ▮ ▮ I ▮ N ▮ L
Row 15: N E X T D O O R ▮ I N D E N E

Solution to crossword 15

```
G L U C O S I C ■ D A R W I N
■ I ■ A ■ Y ■ H ■ D ■ E ■ C ■
O V U M ■ M Y A S T H E N I A
■ E ■ U M ■ R ■ ■ X ■ N ■
W R A S S E ■ L A G R A N G E
■ P ■ S T ■ E E ■ M ■
L O L L O R O S S O ■ I R A N
■ O A ■ I ■ R N ■ C ■
S L A V ■ C A R A G E E N A N
■ ■ O ■ A U E ■ D ■
A N T I L L E S ■ G E I G E R
■ A S ■ T A R M ■
O S M I R I D I U M ■ A C I D
■ A E V N O Q C ■
F L O R E Y ■ G A W K I E S T
```

Solution to crossword 16

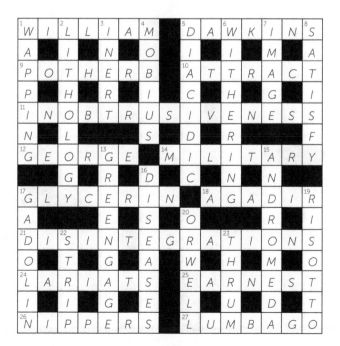

```
W I L L I A M ■ D A W K I N S
A ■ I ■ N ■ O ■ I ■ I ■ M ■ A
P O T H E R B ■ A T T R A C T
P ■ H ■ R ■ I ■ C ■ H ■ G ■ I
I N O B T R U S I V E N E S S
N ■ L ■ S ■ D ■ R ■ F
G E O R G E ■ M I L I T A R Y
■ G ■ R ■ D C ■ N ■ N
G L Y C E R I N ■ A G A D I R
A ■ E ■ S ■ O ■ R ■ I
D I S I N T E G R A T I O N S
O ■ T ■ G A W ■ H ■ M ■ O
L A R I A T S ■ E A R N E S T
I ■ I ■ G E L ■ U ■ D ■ T
N I P P E R S ■ L U M B A G O
```

Solution to crossword 17

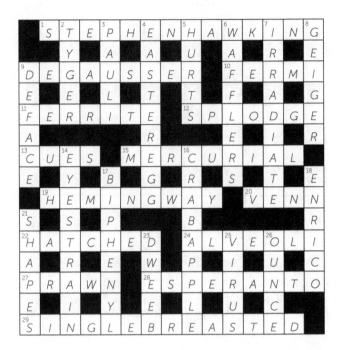

Solution to crossword 18

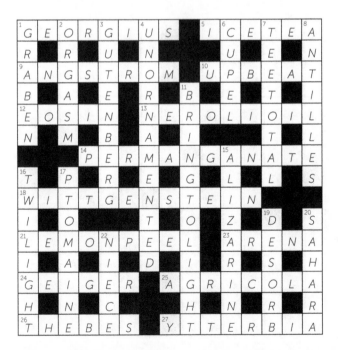

Solution to crossword 19

1	2	3	4	5		6	7	8

Row 1: D J E R A S S I ▓ D O S A G E
Row 2: ▓ O ▓ A ▓ A ▓ N ▓ R ▓ P ▓ U ▓
Row 3: G L Y C O L ▓ F R E D E R I C
Row 4: ▓ I ▓ E ▓ V ▓ I ▓ S ▓ C ▓ D ▓
Row 5: W O O D W A R D ▓ S H I R A Z
Row 6: ▓ T ▓ O ▓ R ▓ E ▓ M ▓ N ▓
Row 7: ▓ ▓ O B S O L E S C E N C E
Row 8: ▓ B ▓ S ▓ A ▓ T ▓ N ▓ E ▓
Row 9: H U M M I N G B I R D S
Row 10: ▓ D ▓ O ▓ E ▓ O ▓ G
Row 11: M A M M A L ▓ D E N A R I U S
Row 12: ▓ P ▓ E ▓ O ▓ S ▓ T ▓ A ▓ I ▓
Row 13: N E U T R I N O ▓ I O D I N E
Row 14: ▓ S ▓ E ▓ R ▓ R ▓ U ▓ O ▓ E ▓
Row 15: S T A R V E ▓ E M M E N T A L

Solution to crossword 20

Row 1: P U N C H E D ▓ P R O C E S S
Row 2: E ▓ O ▓ A ▓ A ▓ O ▓ N ▓ L ▓ A
Row 3: R I N G B I T ▓ L I T H I U M
Row 4: I ▓ L ▓ E ▓ U ▓ O ▓ H ▓ T ▓ O
Row 5: D M I T R I M E N D E L E E V
Row 6: O ▓ N ▓ ▓ S ▓ I ▓ S ▓ ▓ A
Row 7: T H E O R Y ▓ C U L P E P E R
Row 8: ▓ A ▓ E ▓ A ▓ M ▓ O ▓ R ▓
Row 9: T E R R A I N S ▓ S T R I N G
Row 10: H ▓ ▓ C ▓ D ▓ S ▓ ▓ E ▓ E
Row 11: E L E C T R O P H O R E S I S
Row 12: R ▓ T ▓ A ▓ R ▓ I ▓ U ▓ T ▓ T
Row 13: A T H A N O R ▓ S O M A L I A
Row 14: P ▓ Y ▓ T ▓ A ▓ H ▓ B ▓ E ▓ L
Row 15: Y E L T S I N ▓ A N A L Y S T

Solution to crossword 21

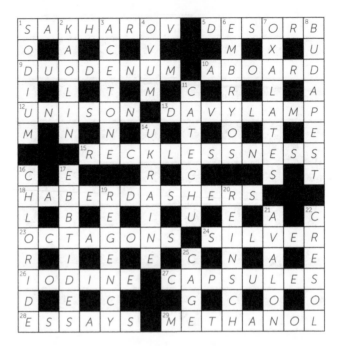

Solution to crossword 22

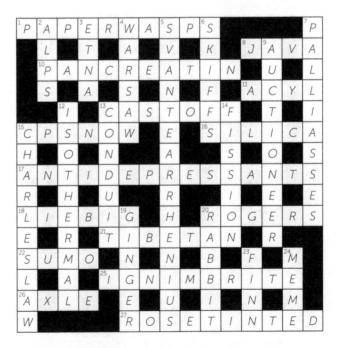

Solution to crossword 23

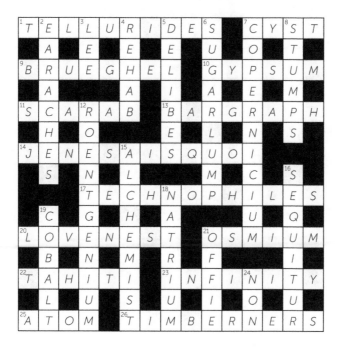

Solution to crossword 24

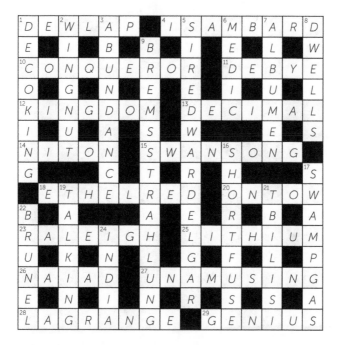

Solution to crossword 25

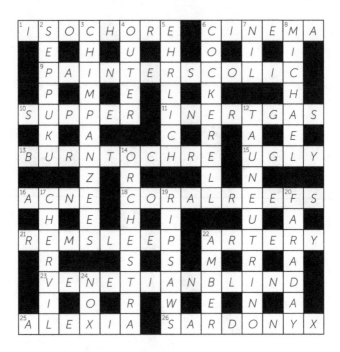

Solution to crossword 26

Solution to crossword 27

F	I	S	H	S	K	I	N	D	I	S	E	A	S	E
R		K		C		L		O		I		P		X
A	M	I	S	H		L	A	G	E	R	L	O	U	T
P		V		R		U		F		T		T		A
P	A	V	L	O	V	S		P	Y	R	R	H	I	C
E		Y		D		T		O		A		E		I
			C	I	T	R	A	L		N	I	G	H	T
S		M		N		A	A		K		M		O
W	R	O	N	G		T	H	R	O	W	N			
A		N		E		E		I		H		A		E
M	E	G	A	R	A	D		M	A	I	L	L	O	T
P		O		S			E		T		K		C
G	O	L	D	C	R	E	S	T		T	E	A	C	H
A		I		A		R		E		L		L		E
S	V	A	N	T	E	A	R	R	H	E	N	I	U	S

Solution to crossword 28

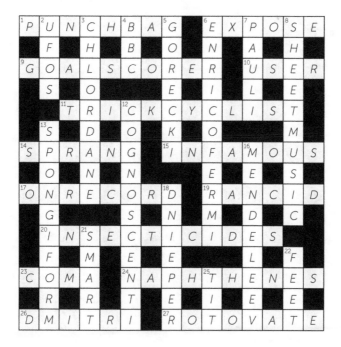

P	U	N	C	H	B	A	G		E	X	P	O	S	E
	F		H		B		O		N		A		H	
G	O	A	L	S	C	O	R	E	R		U	S	E	R
	S		O			E		I		L		E	
		T	R	I	C	K	C	Y	C	L	I	S	T
	S		D		O		K	O				M	
S	P	R	A	N	G		I	N	F	A	M	O	U	S
	O		N		N			E		E		S	
O	N	R	E	C	O	R	D		R	A	N	C	I	D
	G			S		N		M		D		C	
	I	N	S	E	C	T	I	C	I	D	E	S	
	F		M		E		E			L		F	
C	O	M	A		N	A	P	H	T	H	E	N	E	S
	R		R		T		E		I		E		E	
D	M	I	T	R	I		R	O	T	O	V	A	T	E

Solution to crossword 29

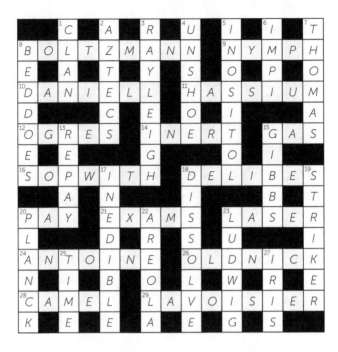

Solution to crossword 30

Solution to crossword 31

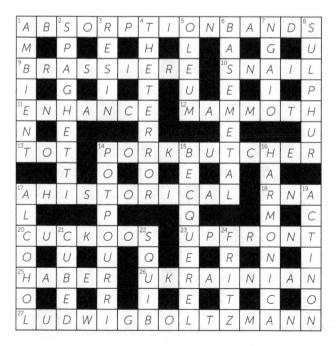

1O	F	2F	3S	H	4O	R	E	■	5A	6G	A	7P	E	8S
N	■	O	■	A	■	A	■	■	A	■	O	■	N
9E	M	P	I	R	I	C	S	■	10E	L	N	I	N	O
M	■	■	I	■	H	■	11S	■	V	■	N	■	W
12A	G	13R	O	C	H	E	M	I	C	A	L	S
N	■	A	■	O	■	L	■	L	■	N	■	E	■	14H
15B	I	Z	E	T	■	16C	H	E	M	I	S	T	R	Y
A	■	O	■	■	A	■	N	■	■	T	■	G
17N	O	R	W	18A	Y	R	A	T	■	19F	R	I	A	R
D	■	B	■	S	■	S	■	S	■	O	■	A	■	O
■	20L	E	B	T	O	S	P	I	R	O	S	I	S
21P	■	A	■	I	■	N	■	R	■	T	■	■	C
22A	D	D	E	R	S	■	23L	I	B	R	E	24T	T	O
L	■	E	■	I	■	■	N	■	A	■	A	■	P
25M	E	S	O	N	S	■	26A	G	A	N	I	P	P	E

Solution to crossword 32

1A	B	2S	O	3R	P	4T	I	5O	N	6B	A	N	D	8S
M	■	P	■	E	■	H	■	L	■	A	■	G	■	U
9B	R	A	S	S	I	E	R	E	■	10S	N	A	I	L
I	■	G	■	I	■	T	■	U	■	E	■	I	■	P
11E	N	H	A	N	C	E	■	12M	A	M	M	O	T	H
N	■	E	■	■	R	■	■	E	■	■	■	U
13T	O	T	■	14P	O	R	K	15B	U	T	C	16H	E	R
■	■	T	■	O	■	O	■	E	■	A	■	A
17A	H	I	S	T	O	R	I	C	A	L	■	18R	N	19A
L	■	■	P	■	■	Q	■	■	M	■	C
20C	U	21C	K	O	O	22S	■	23U	P	24F	R	O	N	T
O	■	U	■	U	■	Q	■	E	■	R	■	I
25H	A	B	E	R	■	26U	K	R	A	I	N	I	A	N
O	■	E	■	R	■	I	■	E	■	T	■	C	■	O
27L	U	D	W	I	G	B	O	L	T	Z	M	A	N	N

Solution to crossword 33

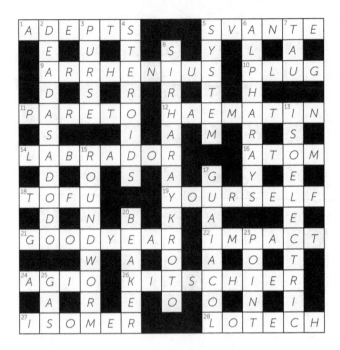

Solution to crossword 34

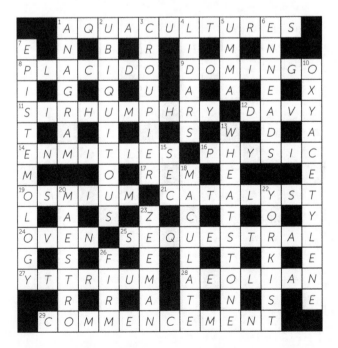

Solution to crossword 35

O	S	I	R	I	S	■	L	I	G	H	T	E	S	T
■	W	■	E	■	E	■	I	■	O	■	U	■	P	■
P	E	N	C	I	L	L	E	A	D	■	R	O	U	X
■	D	■	O	■	F	■	B	■	■	■	N	■	T	■
■	I	N	V	E	S	T	I	G	A	T	I	O	N	S
■	S	■	E	■	E	■	T	■	M	■	N	■	I	■
T	H	O	R	I	A	■	C	A	P	R	O	C	K	S
I	■	■	E	■	R	■	O	■	H	■	N	■	■	O
C	R	E	D	I	T	O	N	■	E	R	E	B	U	S
■	A	■	M	■	E	■	D	■	T	■	S	■	B	■
B	L	U	E	G	R	E	E	N	A	L	G	A	E	■
■	E	■	M	■	■	■	N	■	M	■	R	■	R	■
K	I	L	O	■	F	A	S	C	I	N	A	T	O	R
■	G	■	R	■	A	■	E	■	N	■	V	■	U	■
R	H	A	Y	A	D	E	R	■	E	V	E	N	S	O

Solution to crossword 36

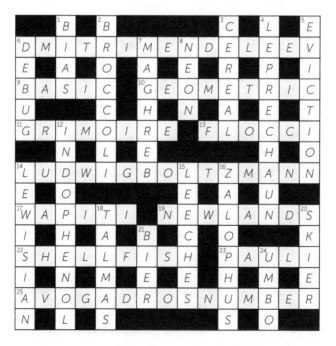

Solution to crossword 37

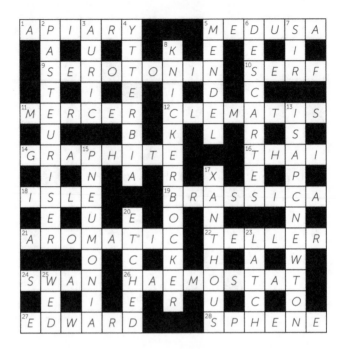

Solution to crossword 38

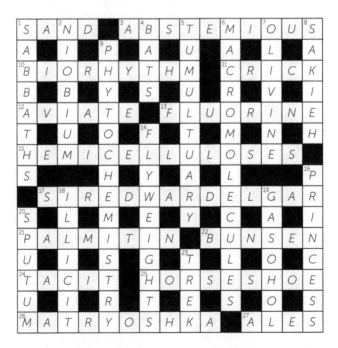

Solution to crossword 39

```
C I C A D A S   ■ M A J O R C A
O ■ L ■ M ■ I ■ E ■ O ■ U ■ L
C R U C I B L E S ■ S U S H I
A ■ E ■ T ■ I ■ O ■ E ■ S ■ M
■ ■ A R A C H N O P H O B E
C ■ A ■ I ■ O ■ S ■ H ■ P ■ N
H O F F M A N N ■ U P S H O T
R ■ T ■ E ■ E ■ C ■ R ■ I ■ A
I L E A N A ■ T H R I L L E R
S ■ R ■ D ■ D ■ A ■ E ■ E ■ Y
T A S T E F U L N E S S ■
I ■ H ■ L ■ C ■ U ■ T ■ P ■ I
A W A K E ■ A L K A L O I D S
A ■ F ■ E ■ T ■ A ■ E ■ K ■ I
N A T I V E S ■ H U Y G E N S
```

Solution to crossword 40

```
■ B I C A R B O N A T E S ■
G ■ E ■ H ■ E ■ E ■ L ■ N
R I C H A R D ■ R U E I N G ■ S
A ■ Q ■ D ■ R ■ S ■ X ■ U ■ U
P A U L ■ A U S T R A L I A N
H ■ E ■ ■ T ■ E ■ N
E H R L I C H ■ D I D E R O T
N ■ E ■ N ■ ■ E ■ I ■ U
E N L A C E D ■ B O R O D I N
■ ■ U ■ A ■ U ■ ■ G ■ G
M I D D L E W I C H ■ Z E U S
H ■ E ■ C ■ K ■ H ■ L ■ P ■ T
O R I G A M I ■ N A I R O B I
■ ■ S ■ T ■ N ■ E ■ M ■ L ■ C
I M M E A S U R A B L E
```

Solution to crossword 41

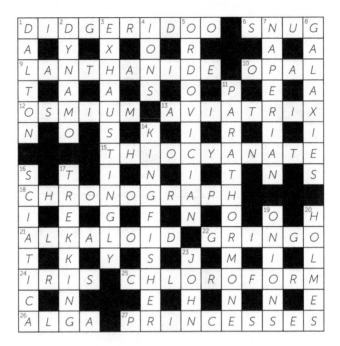

Solution to crossword 42

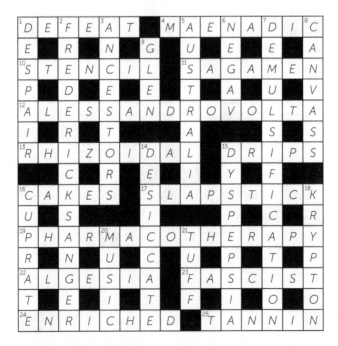

Solution to crossword 43

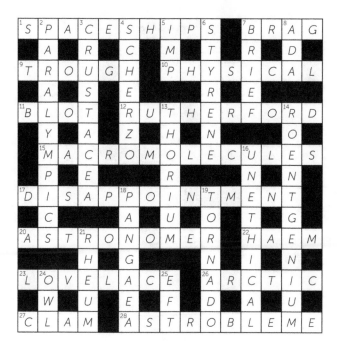

Solution to crossword 44

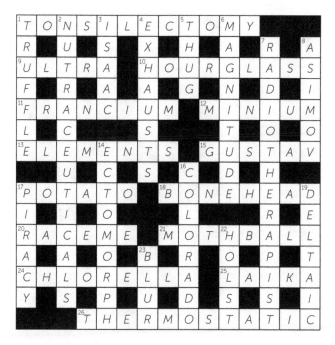

Solution to crossword 45

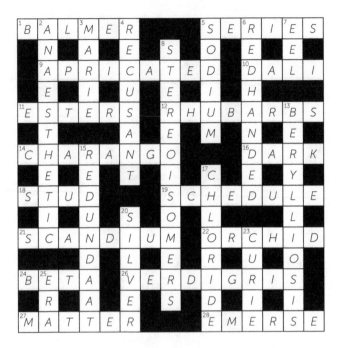

Solution to crossword 46

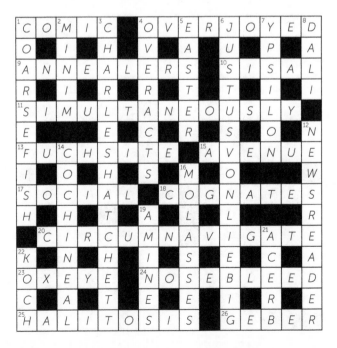

Solution to crossword 47

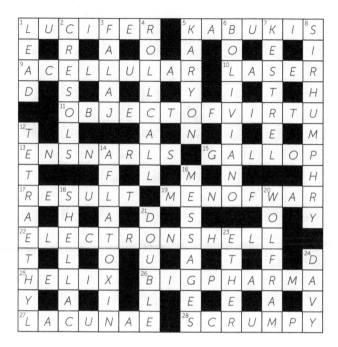

Across/Down grid solution:

1 L	U	2 C	I	3 F	E	4 R	■	5 K	A	6 B	U	K	7 I	8 S
E	■	R	■	A	■	O	■	A	■	O	■	E	■	I
9 A	C	E	L	L	U	L	A	R	■	10 L	A	S	E	R
D	■	S	■	A	■	L	■	Y	■	I	■	T	■	H
■	11 O	B	J	E	C	T	O	F	V	I	R	T	U	
12 T	■	L	■	■	■	A	■	N	■	I	■	E	■	M
13 E	N	S	N	14 A	R	L	S	■	15 G	A	L	L	O	P
T	■	■	■	F	■	L	■	16 M	■	N	■	■	■	H
17 R	E	18 S	U	L	T	■	19 M	E	N	O	F	20 W	A	R
A	■	H	■	A	■	21 D	■	S	■	■	■	O	■	Y
22 E	L	E	C	T	R	O	N	S	H	23 E	L	L	■	
T	■	L	■	O	■	U	■	A	■	T	■	F	■	24 D
25 H	E	L	I	X	■	26 B	I	G	P	H	A	R	M	A
Y	■	A	■	I	■	L	■	E	■	E	■	A	■	V
27 L	A	C	U	N	A	E	■	28 S	C	R	U	M	P	Y

Solution to crossword 48

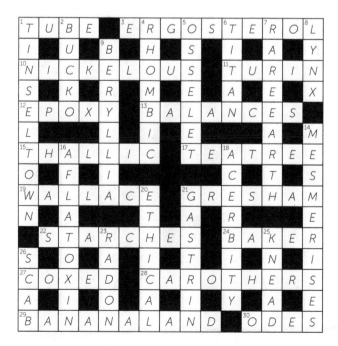

1 T	U	2 B	E	■	3 E	4 R	G	5 O	S	6 T	E	7 R	O	8 L
I	■	U	■	9 B	■	H	■	S	■	I	■	A	■	Y
10 N	I	C	K	E	L	O	U	S	■	11 T	U	R	I	N
S	■	K	■	R	■	M	■	E	■	A	■	E	■	X
12 E	P	O	X	Y	■	13 B	A	L	A	N	C	E	S	■
L	■	■	■	L	■	I	■	E	■	■	■	A	■	14 M
15 T	H	16 A	L	L	I	C	■	17 T	E	18 A	T	R	E	E
O	■	F	■	I	■	■	■	■	■	C	■	T	■	S
19 W	A	L	L	A	C	20 E	■	21 G	R	E	S	H	A	M
N	■	A	■	■	■	T	■	A	■	R	■	■	■	E
■	22 S	T	A	23 R	C	H	E	S	■	24 B	A	25 K	E	R
26 S	■	O	■	A	■	I	■	T	■	I	■	N	■	I
27 C	O	X	E	D	■	28 C	A	R	O	T	H	E	R	S
A	■	I	■	O	■	A	■	I	■	Y	■	A	■	E
29 B	A	N	A	N	A	L	A	N	D	■	30 O	D	E	S

Solution to crossword 49

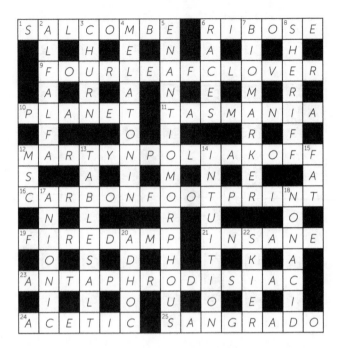

Solution to crossword 50

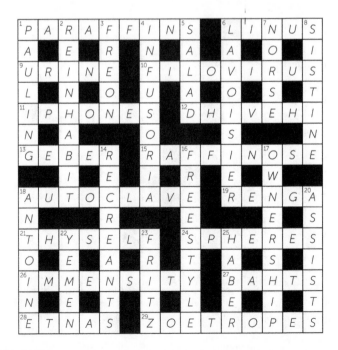

Solution to crossword 51

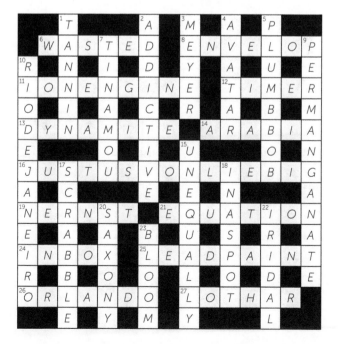

Solution to crossword 52

Solution to crossword 53

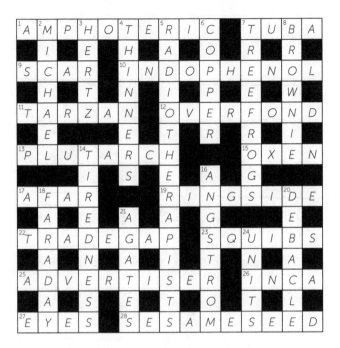

Solution to crossword 54

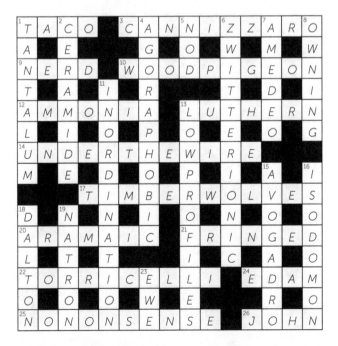

Solution to crossword 55

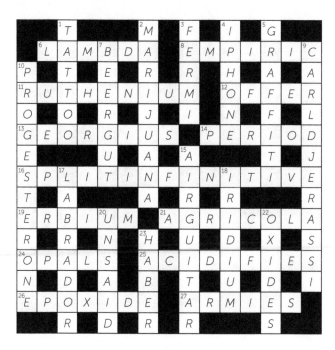

Solution to crossword 56

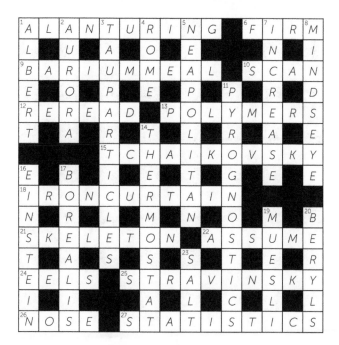

Solution to crossword 57

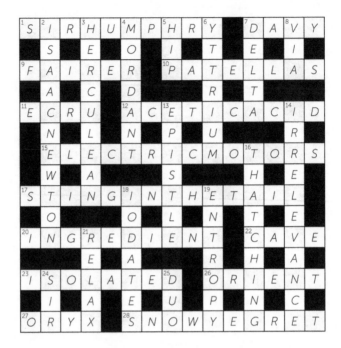

Solution to crossword 58

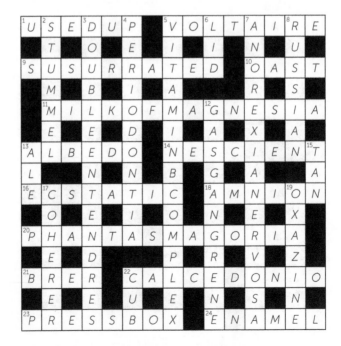

Solution to crossword 59

Solution to crossword 60

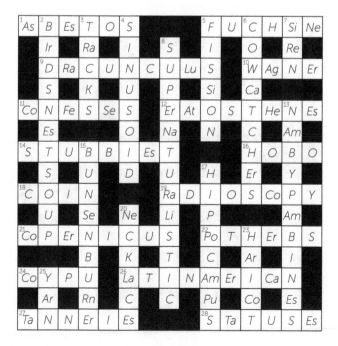

Solution to crossword 61

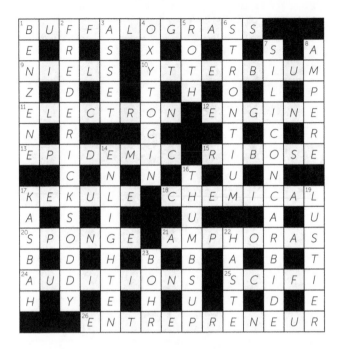

Solution to crossword 62

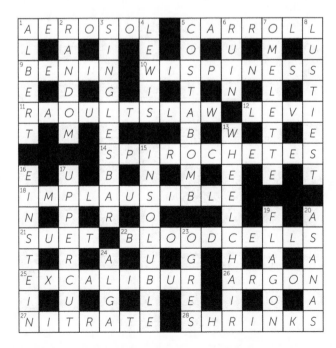

Solution to crossword 63

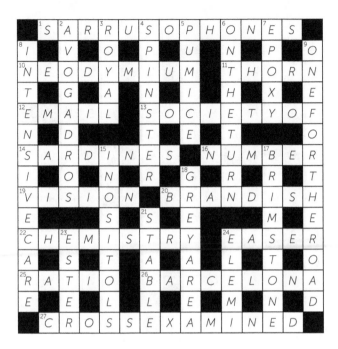

Solution to crossword 64

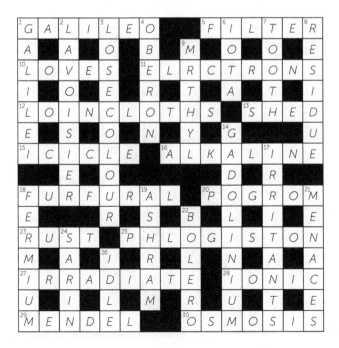

Solution to crossword 65

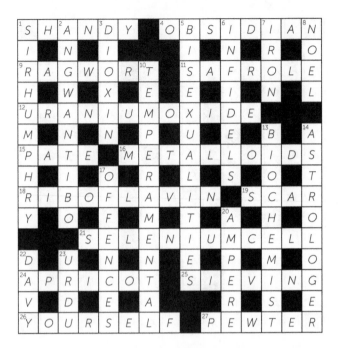

S	H	A	N	D	Y		O	B	S	I	D	I	A	N
I		N		I			I		N		R			O
R	A	G	W	O	R	T	S	A	F	R	O	L	E
H		W		X		E		E		I		N		L
U	R	A	N	I	U	M	O	X	I	D	E			
M		N		N		P		U		E		B		A
P	A	T	E		M	E	T	A	L	L	O	I	D	S
H		I		O		R		L		S		O		T
R	I	B	O	F	L	A	V	I	N		S	C	A	R
Y		O		F		M		T		A		H		O
			S	E	L	E	N	I	U	M	C	E	L	L
D		U		N		N		E		P		M		O
A	P	R	I	C	O	T		S	I	E	V	I	N	G
V		D		E		A			R			S		E
Y	O	U	R	S	E	L	F		P	E	W	T	E	R

Solution to crossword 66

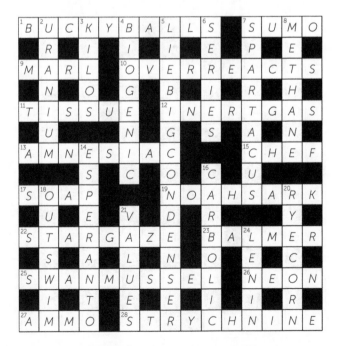

B	U	C	K	Y	B	A	L	L	S		S	U	M	O
	R		I		I		I		E		P		E	
M	A	R	L		O	V	E	R	R	E	A	C	T	S
	N		O		G		B		I		R		H	
T	I	S	S	U	E		I	N	E	R	T	G	A	S
	U			N			G		S		A		N	
A	M	N	E	S	I	A	C			C	H	E	F	
			S		C		O		C		U			
S	O	A	P			N	O	A	H	S	A	R	K	
	U		E		V		D		R				Y	
S	T	A	R	G	A	Z	E		B	A	L	M	E	R
	S		A		L		N		O		E		C	
S	W	A	N	M	U	S	S	E	L		N	E	O	N
	I		T		E		E		I		I		R	
A	M	M	O		S	T	R	Y	C	H	N	I	N	E

Solution to crossword 67

Solution to crossword 68

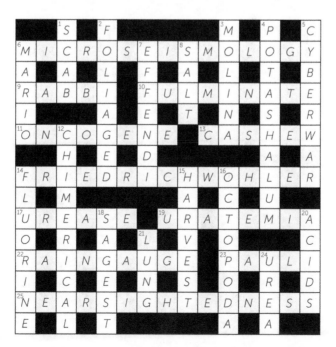

Solution to crossword 69

Solution to crossword 70

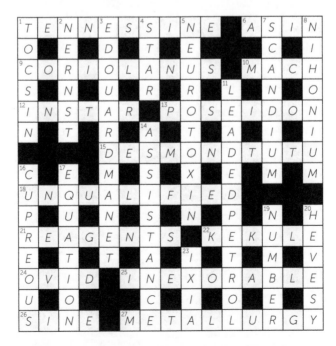